THE ROAD TO
SCIENTIFIC
SUCCESS

Inspiring Life Stories of Prominent Researchers

Volume 1

THE ROAD TO
SCIENTIFIC
SUCCESS

Inspiring Life Stories of Prominent Researchers

Volume 1

editor

Deborah D L Chung

University at Buffalo, State University of New York, USA

World Scientific

NEW JERSEY • LONDON • SINGAPORE • BEIJING • SHANGHAI • HONG KONG • TAIPEI • CHENNAI

36 8632

Published by

World Scientific Publishing Co. Pte. Ltd.
5 Toh Tuck Link, Singapore 596224
USA office: 27 Warren Street, Suite 401-402, Hackensack, NJ 07601
UK office: 57 Shelton Street, Covent Garden, London WC2H 9HE

British Library Cataloguing-in-Publication Data
A catalogue record for this book is available from the British Library.

THE ROAD TO SCIENTIFIC SUCCESS:
Inspiring Life Stories of Prominent Researchers (Volume 1)

ISBN 981-256-600-7
ISBN 981-256-466-7 (pbk)
ISSN 1793-2823

Printed by FuIsland Offset Printing (S) Pte Ltd, Singapore

Preface

This is the inaugural volume of a new book series, titled "The Road to Scientific Success: Inspiring Life Stories of Prominent Researchers". It describes the road to scientific success, as experienced and described by prominent researchers. The focus on research process (rather than research findings) and on personal experience is intended to encourage the readers, who will be inspired to be dedicated and effective researchers.

The objectives of this book series include the following.

1. To motivate young people to pursue their vocations with rigor, perseverance and direction
2. To inspire students to pursue science or engineering
3. To enhance the scientific knowledge of students that do not major in science or engineering
4. To help parents and teachers prepare the next generation of scientists or engineers
5. To increase the awareness of the general public to the advances of science
6. To provide a record of the history of science

The book series differs from existing publications in its emphasis on the personal experience of prominent scientific researchers. In contrast, existing books mostly address either scientific research findings or general research methodology.

This book series also differs from existing publications that provide life stories that are not written by the people in the stories. In contrast, each story in this book series is a first-hand description by the person involved.

Due to the historic significance of the life stories and the impact of the scientific advances behind each story, this book is expected to be valuable from the viewpoint of the history of science. The uniqueness of the book series relates to the following.

1. Untold life stories in the words and photos of world-class scientists
2. Keys to scientific success elucidated
3. Scientific research strategies disclosed
4. Career preparation methods demonstrated
5. History of science revealed
6. Up-to-date science covered in a way that readers with little or no science background can understand

This book series is suitable for use in courses on introduction to science or engineering, professional success methodology, scientific research methodology and history of science. Suitable readers include university students (undergraduate and graduate levels), college students, high school students, educators, teachers, education workers/ administrators, historians, technology business personnel, technology transfer professionals, research managers, scientists, engineers, technicians, parents and the general public.

It is recommended that teachers use selected chapters as required reading for secondary school, college or university students, regardless of the intended major of the students, and have students turn in homework in the form of essays on the impact felt and lessons learned after the reading.

Deborah D.L. Chung
Editor
Buffalo, NY, U.S.A.
Sept. 2005

Editor: Deborah D.L. Chung, Endowed Chair Professor,
University at Buffalo, State University of New York, USA
(E-mail: ddlchung@buffalo.edu;
http://www.wings.buffalo.edu/academic/department/eng/mae/cmrl)

Contents

Chapter 1

Deborah Duen Ling Chung: Innovator in Engineering Materials Use

1.1 Introduction by the Editor

1.1.1 *Use of engineering materials*

Engineering materials are solids that are relevant to engineering applications. They are needed for any product and are considered to be the foundation of technology. They include metals, ceramics, organic materials (e.g., polymers), semiconductors and composites. Their applications include those related to aerospace, automobile, construction, electronic, energy, environmental and biomedical industries.

The value of any material depends on the ability of the material to satisfy one or more needs of technology and the ability to compete in performance and cost with other materials that may also satisfy these needs. Without a viable application, a material cannot attain commercial importance. For a new material, application development is obviously necessary. Successful development of a new use for an existing material means the opening of a new market for the material. Therefore, for both new and existing materials, the development of applications is critical to attaining and maintaining commercial value.

Due to the wide variety of industries that use materials, the development of material applications requires broad knowledge of the current needs of various industries. Due to the range of materials that may be available for a given application, comparative testing of a

1

number of materials is usually necessary in order to find out how well a material competes with other materials. Thus, broad knowledge of various types of materials is also needed for the development of material applications.

1.1.2 *Scientific contributions of Dr. Chung*

Dr. Chung's materials research has resulted in a number of innovative uses of materials. They include (i) the revolutionary use of concrete for sensing loads (as needed for traffic monitoring, room occupancy monitoring and weighing), (ii) the use of carbon materials for electromagnetic interference (EMI) shielding (which is needed for protecting electronics from interference by radio waves, such as those emitted by cellular phones), (iii) the use of carbon black paste for improving thermal contacts (e.g., improving the contact between the heat sink and the microprocessor of a computer for the purpose of microelectronic cooling), (iv) the use of the interface between plies of carbon fibers in a polymer-matrix structural composite as a sensor of temperature, impact, damage and their spatial distributions (as needed for aircraft, automobile, sporting goods and other products that use the composite for their structures), (v) the use of the junction of dissimilar cement-based materials or that of dissimilar carbon fiber polymer-matrix composites as a thermocouple junction for sensing temperature, (vi) the use of concrete to suppress its own vibrations, and (vii) the use of titanium in place of glass as a binder in electrically conductive thick-film pastes for electrical interconnections in microelectronics.

In addition to the development of material applications, Dr. Chung has developed (i) electrical methods for studying the interfaces in composite materials and for evaluating joints (e.g., joints obtained by fastening, adhesion or soldering), (ii) methods for making various composite materials and carbon materials, and (iii) understanding of various material phenomena.

Dr. Chung has been active in disseminating knowledge and promoting scholarship through the authoring of textbooks and reference books, the authoring of 40 review papers and 400 research papers that have been published in archival professional journals, the authoring of 5 encyclopedia articles, the giving of 200 invited lectures in professional conferences, universities and industries, interviews by the mass media (CNN, NBC, etc.), the teaching of materials science to over 3,000 students, and serving as the major professor of 29 Ph.D. graduates.

1.1.3 *Honors received by Dr. Chung*

Dr. Chung received the Charles E. Pettinos Award from the American Carbon Society in 2004. This is a triennial international award to recognize one person or one research group for outstanding recent research accomplishments in carbon science and technology. Dr. Chung is the first woman in America, the second woman in the world, and the first person of Chinese descent to receive this honor. In addition, Dr. Chung received the Hsun Lee Award from Institute of Metal Research (Chinese Academy of Sciences) in 2005. It is an award to recognize research accomplishments in materials science and technology. Furthermore, she received the Chancellor's Award for Excellence in Scholarship and Creative Activities in 2003 from the State University of New York and the Outstanding Inventor Award in 2002 (Fig. 1.1) from the same university system. This university system consists of over 60 campuses. Dr. Chung also received the Hardy Gold Medal from American Institute of Mining, Metallurgical, and Petroleum Engineers in 1980, and the Ladd Award from Carnegie-Mellon University in 1979. Dr. Chung is a Fellow of American Carbon Society (since 2001) and a Fellow of ASM International (since 1998, Fig. 1.2). Other awards include the Teetor Educational Award from Society of Automotive Engineers in 1987 and the Teacher of the Year Award from Tau Beta Pi in 1993.

1.1.4 *Career development of Dr. Chung*

Dr. Chung received her B.S. (Honor) degree in Engineering and Applied Science and M.S. degree in Engineering Science from California Institute of Technology (abbreviated Caltech) in 1973. After that, she went on to Massachusetts Institute of Technology (MIT), where she

Fig. 1.1 Dr. Chung receiving the Outstanding Inventor Award in 2002 from Chancellor King (right of photo) of the State University of New York system.

Fig. 1.2 Dr. Chung becoming Fellow of ASM International in 1998.

received S.M. degree in Materials Science in 1975 and Ph.D. degree in Materials Science in 1977.

Since 1977, Dr. Chung has been a professor, first at Carnegie-Mellon University (1977-1986) and then at State University of New York at

Buffalo (1986-present). She moved up in academic rank step by step – assistant professor (1977-1982), associate professor (1982-1986), professor (1986-present) and endowed chair professor (1991-present). In addition to her full-time professorship in U.S.A., Dr. Chung holds honorary professorships in a number of universities in China, including Tianjin University, Wuhan University of Technology (Fig. 1.3), Southeast University, Harbin Institute of Technology, Shantou University, Jinan University and Beijing Technology and Business University.

Dr. Chung is also a pianist, holding the L.R.S.M. diploma in piano performance (1971) from the Royal Schools of Music, UK. She has performed in various countries, including USA and China. Her repertoire includes her own compositions.

1.2 Dr. Chung's Description of Her Life Experience

1.2.1 *In school in Hong Kong*

I was born and raised (Fig. 1.4) in Hong Kong (Fig. 1.5), when it was a British colony. Thus, I was born British, though I am of Chinese descent. My maternal grandfather (Fig. 1.6) was a revolutionary. The revolution, led by Dr. Sun Yat-Sen, turned China from an empire (Qing Dynasty) to a republic.

Due to colonialism, career advancement for local Chinese people was quite limited. This seriously affected the professional advancement of my parents. My father, Leslie Wah-Leung Chung, who had been wounded in active combat in Hong Kong during the Japanese invasion in the Second World War, taught management in the post-secondary level (Fig. 1.7(a)). My mother, Rebecca Chan Chung, who was a nurse with the Flying Tigers in Kunming, China, during the same war, taught nursing in a hospital system in Hong Kong (Fig. 1.7(b)). I often heard from my father about the unfair treatments that he and his colleagues faced at work due to racial bias.

Fig. 1.3 Dr. Chung lecturing in Wuhan University of Technology, China, in 2002.

Fig. 1.4 Deborah Chung with her extended family in Hong Kong at the age of 3 months.
She was held by her mother, who was next to her father.

Fig. 1.5 Hong Kong around 1960.

Fig. 1.6 Dr. Chung's maternal grandparents (Mr. and Mrs. Bo Yin Chan) and mother
(baby).

I have always been afraid of dogs. Once, when I was walking outdoor in
Hong Kong, I was scared by the sudden appearance of a dog, which was
walked by a British caucasian lady. Due to the scare, I dodged aside.
Instead of apologizing to me for causing the scare, the lady expressed
dissatisfaction of how I treated her dog. I felt hurt, as the dog was
considered more important than me. I held a British passport, but the
passport stated that I could not live in the United Kingdom – a clear
indication of a second-class citizenship.

In spite of the discrimination, I am grateful for the school system
established by the British people. I attended Ying Wa Girl's School
("Ying" meaning British; "Wa" meaning Chinese) for both my primary
and secondary school education. This school was founded in 1900 by
the London Missionary Society, which was the Christian mission
organization (founded in 1795) that sent the first Protestant missionary,
Robert Morrision (1782-1834), to China. Morrison arrived in China in
1807 and translated the Bible to Chinese. The missionaries, all British,
that taught in the school showed genuine and sacrificial love – not a bit
of discrimination. The school principal, Miss Silcocks (a missionary,
Fig. 1.8) was highly respected and dearly loved by the students. Most of
the missionaries spoke Cantonese Chinese well. Each morning, the

entire student body gathered together for about 20 minutes to sing hymns, pray and hear about the principles and meaning of life. It was wonderful to receive spiritual nourishment in addition to rigorous academic training and education in both Chinese and English. I studied in this school from Grade 3 to Grade 11.

(a)   (b)

Fig. 1.7 (a) Dr. Chung's father, Mr. Leslie Wah-Leung Chung. (b) Dr. Chung's mother, Mrs. Rebecca Chan Chung, working as a nurse in China during the Second World War.

Fig. 1.8 Miss Silcocks, principal of Ying Wa Girls School (1960's)

I am grateful to my parents, my maternal grandmother, and my school teachers (particularly Ms. Sui Mooi Chow of Ying Wa Girls' School), who taught me since I was a toddler the importance of honesty, diligence, and having a purpose in life that is rooted in God. I started taking piano lessons at the age of five. Even at that young age, I spent an hour a day practicing at the piano. I remember putting a clock at the piano and continuing with the practice until the hour was up. During the summer vacations, I made a time table for myself so as to make good use of my time for studies and piano practice. This sense of discipline has remained with me throughout my life.

By the time I was near graduation from secondary school, I had become quite good at the piano. I seriously considered making music my career. However, my father advised me not to do so, because of the extreme competition among performers, the difficulty of becoming a world-class performer and the financial hardship often faced by artists that are not world-class.

While I was in Grade 11 (known as Form 5, the last year of secondary school in the Hong Kong education system), the first person landed on the moon. It was the year 1969. I was so excited by this scientific achievement of the U.S. space program that I told my parents that I would like to go to the U.S. for university education in engineering.

1.2.2 *In university in USA*

In 1970, after completing Grade 12 (the first year of a two-year matriculation course) at King's College, Hong Kong, I went by myself to the US to study. It was very expensive for my parents to support my university education in the U.S., so I studied extremely hard, trying to graduate early, thereby lessening the financial burden of my parents. In just three years, I managed to graduate from California Institute of Technology (Caltech) with both B.S. and M.S. degrees. I also turned out to be one of the four first woman B.S. graduates of Caltech (Fig. 1.9).

My education in Caltech emphasized electrical engineering and computer science. It was inspiring to study under professors of international stature and to learn about the latest technology. I was particularly privileged to learn from Professor Carver Mead (Fig. 1.10)

Fig. 1.9 The four first woman B.S. graduates of California Institute of Technology (1973). Deborah Chung is at the center. (Photo taken by Floyd Clark of Caltech.)

Fig. 1.10 Professor Carver Mead of California Institute of Technology. (Photo taken by Jon Brennis of California Institute of Technology.)

(a)   (b)

Fig. 1.11 (a) Professor Mildred S. Dresselhaus of Massachusetts Institute of Technology. (b) Dresselhaus and Chung on the occasion of Chung's Pettinos Award ceremony in 2004.

about the design and fabrication of integrated circuits. To do so in the early 1970's was much ahead of its time.

The most precious aspect of my Caltech undergraduate education was research participation. Due to the small size of the student body (just about 200 students per class, i.e., about 800 undergraduate students in the whole university, in addition to about 800 graduate students), even undergraduate students had ample opportunity to participate in research and get paid for the work. Through research, I got much more excited about studying. I wanted to understand fully what I was learning. Studying thus took on a higher level of rigor. The reading of textbooks alone could not satisfy me, so I read numerous journal papers on my own. The reading was not for any course, but was to satisfy my thirst for knowledge. The research, though low in level due to my limited knowledge, was like a lightning rod to me. I am thankful to Caltech for the opportunity to be exposed to research when I was still an undergraduate student. The financial help provided by the research work was also wonderful.

The research that I conducted at Caltech was under the late Professor Pol Duwez, the father of amorphous metals. Through the research, I was introduced to the subject of materials science. It was a subject that I had not studied before, but I came to like it so much that I decided to pursue Ph.D. education in this area. To me, the attraction of this subject lies in its practicality in addition to its rich science. Materials are needed for any technology and the advent of materials has marked the history of human civilization, as shown by the Stone Age, Bronze Age and Iron Age.

If I were to stay in Caltech, I could have been able to receive a Ph.D. degree in computer science in two years. However, I decided to forsake computer science for materials science. Thus, in 1973, I left Caltech and moved on to M.I.T., which was (and still is) a top university in the field of materials science and engineering.

1.2.3 *Research in graphite*

While I was at M.I.T. as a graduate student in the Department of Materials Science and Engineering, I took a solid-state physics course from Professor Mildred S. Dresselhaus (Fig. 1.11). The beauty of the physics grabbed me and I decided to do research under Professor Dresselhaus.

Professor Dresselhaus had been conducting research on graphite, particularly in relation to its electronic behavior. At that time (1974), she wanted to start research on compounds of graphite, and I became her first student of this area of research. These compounds, called intercalation compounds, are the products of reactions between graphite and very reactive chemicals (such as bromine and nitric acid). Graphite has a crystal structure that consists of layers of carbon atoms, with each layer having a honeycomb structure (Fig. 1.12). Upon reaction, the foreign species resides between the carbon layers, thus resulting in a layered compound. Due to the transfer of electrons between the foreign species and graphite during the reaction, the compounds are much more conductive electrically than graphite itself. In fact, the compounds can become metallic and, in that case, they are called synthetic metals.

Research in the area of graphite intercalation compounds was intriguing due to the richness of its physics and chemistry and the breadth of material properties that are relevant. Because Professor Dresselhaus and I were among the first people in USA to engage in

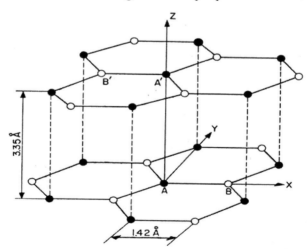

Fig. 1.12 The crystal structure of graphite. Each circle represents a carbon atom.
$1 \text{ Å} = 10^{-10} \text{ m}$.

research in this area, we became recognized internationally for the research in just a few years.

After graduating from M.I.T. with a Ph.D. degree in Materials Science in 1977, I continued research in the area of graphite intercalation compounds while I was an assistant professor (and later an associate professor) in Carnegie-Mellon University, Pittsburgh, Pennsylvania, USA. My research was funded by U.S. National Science Foundation, U.S. Air Force Office of Scientific Research and other government agencies. The research resulted in the understanding of (i) the crystal structure of graphite intercalation compounds, (ii) the process of intercalation compound formation and (iii) the process of exfoliation of an intercalation compound.

Exfoliation is the puffing-up (expansion) of an intercalation compound in the direction perpendicular to the carbon layers, thus resulting in a structure that resembles an accordion. Compression of a collection of such structures without a binder leads to mechanical interlocking of the accordion structures. The interlocking results in the formation of a flexible sheet that is known as "flexible graphite". Because of its resiliency in the direction perpendicular to the sheet, flexible graphite is used as a gasket for fluid sealing. This gasket material is attractive for its ability to resist harsh chemicals and high temperatures. It is a replacement for asbestos, which is a gasket material that suffers from its carcinogenic (cancer causing) character. Flexible graphite is a material developed by Union Carbide Corp. My research at that time addressed the process of exfoliation rather than flexible graphite, though exfoliation is a process that is central to the making of flexible graphite.

By 1985, it was clear that the field of graphite intercalation compounds was declining. The decline was due to the fact that the electrical conductivity attained by intercalation was below that of copper, a widely available electrical conductor. Although the science remained interesting, the research field could not be sustained without a viable application. Thus, research funding in this area diminished and I learned in a hard way the importance of materials applications.

1.2.4 *Research in smart concrete*

Due to the decline in the field of graphite intercalation compounds, I decided to move to other fields. The move was like starting from scratch, as the field of graphite intercalation compounds was my only field of expertise. I was actually a full professor by 1986 and I did not have to move to other fields. However, I decided to take on the challenge by retooling myself.

Unexpectedly, I got interested in concrete. I had had no background on concrete at all. I had not even mixed concrete before. The field of concrete was (and still is) dominated by people with expertise in mechanics, due to the use of concrete in structures and the importance of mechanical analysis of structures. However, my background on mechanics was limited. Thus, I learned from the beginning and started as a "freshman" in the field of concrete.

Although the experts in concrete kindly suggested that I study the mechanical behavior of concrete, I proceeded to study concrete as if it were an electronic material. It was seemingly a crazy direction of research. However, it was a direction that reflected my training in the electronic behavior of materials. I also applied my background on graphite by adding carbon fiber to cement. Thus, I discovered that cement containing a small proportion of short carbon fiber changed its electrical resistivity upon deformation. The change was large and reversible. This means that one can measure the electrical resistance of the material and obtain information on the deformation. In other words, the material is a sensor of its own deformation, i.e., self-sensing. Since the deformation is related to the load, the material can serve as a load sensor as well. A load sensor can be viewed as a scale (i.e., a balance) for weighing things.

Concrete had always been considered a structural material only. It was revolutionary to use concrete as a sensor. The use of a structural material to sense itself is highly attractive compared to the use of an embedded or attached device to attain the sensing function. This is because a structural material is low in cost and is durable compared to the devices. Furthermore, the sensing ability provided by a structural material can be throughout a structure, whereas that provided by an embedded or attached device is only for limited spatial extents. In addition, embedded devices cause degradation of the mechanical properties of the structure, due to the devices resulting in regions of mechanical weakness in the structure.

Applications of the self-sensing concrete (also called smart concrete) include (i) the weighing of vehicles (e.g., trucks) while they are moving, for the purpose of traffic monitoring, border monitoring and preventing the damage of highways by heavy trucks, (ii) the weighing of cargo in a cargo yard, (iii) the weighing of a room for the purpose of determining the room occupancy, which is used to control in real time the heating, cooling, ventilation and lighting of the room for the purpose of saving energy, and (iv) the detection of vibrations in a structure for the purpose of vibration suppression and the consequent enhanced safety of structures, as in the case of an earthquake.

1.2.5 *Research in materials for electromagnetic interference shielding*

Electromagnetic interference (EMI) shielding refers to the blocking of electromagnetic radiation, particularly radio waves, which are emitted by cellular phones. The radiation can cause malfunction of computers and other electronics. As our society depends heavily on computers, there is a growing need to protect computers from such interference. Thus, materials for EMI shielding are needed for electronic enclosures and telephone housings. They are also needed for concert halls, where the ringing of cellular phones is annoying during a concert. If the building material blocks the radiation, cellular phones in the building cannot function. Such buildings are called cell-phone proof buildings. EMI shielding is also needed for embassies and military buildings for the purpose of deterring electromagnetic forms of spying. Such spying is possible because radiation is emitted by telephone, fax machines and computers and the analysis of the emitted radiation can yield some information.

Around 1990, I got interested in EMI shielding. Although electrical engineers had long been working on EMI shielding, materials engineers had done little on the subject. The development of materials for EMI shielding requires materials expertise – not just expertise in electromagnetic radiation. Thus, I realized that the field of materials for EMI shielding was a neglected but important area. With cellular phones being so common, the commercial importance of EMI shielding is now (2005) very clear. However, my early start (1990) in this research allowed me to be well positioned for meeting the growing commercial need.

Carbon (or graphite) is a good reflector of a wide spectrum of electromagnetic radiation. By making use of my background in graphite, in addition to my background in electrical engineering, I came to find that flexible graphite (an old material) is an exceptionally good material for EMI shielding. Because of its resiliency, flexible graphite is particularly attractive as a EMI gasket material, which is needed for the electromagnetic sealing of joints in a shielded enclosure.

I also developed nickel nanofiber (also called nickel filament), which is an exceptionally good filler for providing polymer-matrix composites that are effective for shielding. The high effectiveness of nickel nanofiber is due to its small diameter and high electrical conductivity. A small diameter is advantageous because the radiation only penetrates the near surface region of a conductor.

1.2.6 *Research in materials for microelectronic cooling*

Overheating is the most critical problem in the electronic industry, as it limits the further miniaturization, power, performance and reliability of microelectronics. Materials are needed to alleviate this problem. For the purpose of developing such materials, materials experts were needed. However, the participation of materials experts in this field was limited. The field was dominated by electrical engineers. Thus, I came to realize that this was an area that was neglected but practically very important.

Around 1990, with the help of substantial research funding from U.S. Defense Advanced Research Projects Agency, I started work in developing heat sink materials. A heat sink is made of a material of high thermal conductivity, as it is attached to the microprocessor of a computer for the purpose of channeling the heat from the microprocessor to the environment.

Copper is a common material for heat sinks, due to its high thermal conductivity. However, it suffers from a high coefficient of thermal expansion. Due to the low coefficient of thermal expansion of silicon and its ceramic substrate, it is necessary for the heat sink to have a low coefficient of thermal expansion also. Otherwise, thermal stress occurs and the silicon wafer may become warped. Therefore, I put fillers of low thermal expansion (e.g., molybdenum particles) into copper. The difficulty of making the copper-matrix composite lies in the high volume fraction of the filler that is necessary in order to lower the coefficient of thermal expansion adequately. Thus, my graduate student (Pay Yih) and I developed a method of fabricating copper-matrix molybdenum particle composites that exhibit the required low coefficient of thermal expansion.

Around 1998, I came to realize that the most neglected aspect of the microelectronic cooling technology pertained to the improvement of the thermal contact between the heat sink and the heat source (i.e., the microprocessor). A superb heat sink cannot be effectively utilized, unless the thermal contact is good. Thus, I started to switch my attention from the heat sink to the thermal contact.

Improvement of the thermal contact requires a material placed at the interface between the heat sink and the heat source. This material is known as a thermal interface material. It is commonly in the form of a paste, which is known as a thermal paste. It is necessary for the paste to conform to the surface topography of the adjoining surfaces, because no surface is perfectly smooth and the valleys in the surface topography trap air, which is a thermal insulator. Thus, it is important for the thermal paste to displace the air out of the interface. For this purpose, a high level of conformability is necessary for the thermal paste.

Through a comparative experimental study of a number of thermal interface materials, I came to understand that the most important criterion for the performance of a thermal interface material is conformability. If conformability is poor, the thermal interface material is ineffective, even if it has a high thermal conductivity. However, thermal interface materials have been developed in the electronic industry by maximizing the thermal conductivity at the expense of the conformability. For example, a thermal paste is commonly loaded with so much silver powder (silver being a material of high thermal conductivity) that its conformability is sacrificed.

Due to my graphite background, I understood that carbon black is highly compressible (i.e., squishable). The squishability is due to the fact that carbon black is in the form of porous agglomerates of particles. Due to the squishability and the very small size (30 nm) of the particles, carbon black is highly conformable. With this understanding, I developed carbon black paste, which was found to be exceptionally effective as a thermal paste. Carbon black is thermally conductive, but it is not as conductive as silver. Nevertheless, carbon black paste is more effective than silver paste due to its extraordinary conformability. Furthermore, carbon black is much less expensive than silver.

1.2.7 *Outreach as a scientist*

While remaining active in research, I started in 1998 a new line of work as a scientist. This work is outreach to the general public, particularly students in schools and universities. I felt that I should use my experience to inspire people to pursue science and, more importantly, to have direction and meaning in life. I also noticed students in secondary schools in USA shying away from science, thus resulting in decline in the number of students that pursue science or engineering in USA. A similar trend is occurring in Europe, but the situation is different in Asia. My outreach has taken the form of lectures or concert-lectures titled "The Road to Scientific Success", "My Science, Music and Life", etc. (Fig. 1.13). The addition of music performance to a lecture adds inspiration, excitement and entertainment to the outreach. Outreach events have been held in North America, Europe, China, Singapore and Indonesia, with audience size as high as 2,000.

A 10th Grade student wrote after hearing me:

"I always considered science to be serious, rational, and noble. Before today, I did not realize that even science could also be passionate, because it is also a belief, and it demands love from you... The seemingly aged appearance did not cover her young heart for love in life. It is her love for life that has developed such a great character in her; it is her passion that has made her so charismatic. A young heart made her appear so beautiful on the stage."

Fig. 1.13 Dr. Chung performing in a concert-lecture.

Fig. 1.14 Parents, sister (Maureen Chung, front right) and husband (Lan K. Wong, back center) of Dr. Chung (back right) in 1995.

This book series, for which I serve as Editor, allows the use of not just my experience, but also that of numerous prominent scientists, to inspire. It is common for scientists to talk about their research findings, but it is not common for them to talk about their life journey, their research strategies and their keys to success. I expect that some of the life stories will be made into film documentaries for showing on television, so as to reach the general public broadly throughout the world.

1.2.8 *Reflections of my 30-year career*

Reflection of my 30-year career in scientific research shows the importance of solid training, breadth of knowledge, humility to learn continuously, ability to cross disciplinary boundaries, versatility, adaptability, perseverance, diligence, understanding industrial needs and technological limitations, and creativity.

The word "vocation" has a Latin root, which relates to "vocal", meaning calling. There is a call behind one's vocation, which becomes one's mission in life. The choice of a vocation is not just based on interest or financial considerations. Because of the noble character of the call, one should not readily change from one vocation to another when difficulty arises. It is extremely important to have a purpose in life. Otherwise, one would be doing things in a meaningless way.

The numerous turns in my life journey, starting from my schooling in Hong Kong, my lone move from Hong Kong to USA at the age of 18, my change of major from electrical engineering and computer science to materials science, my attending the top technical universities in the world (Caltech and MIT), the decline of my only field of expertise, my move from this field to fields that were foreign to me, my broadening from graphite to numerous types of materials (cement, polymers, ceramics, metals and composites) and from electronic behavior to other types of behavior (mechanical, thermal and chemical behavior), my invention of smart concrete, nickel nanofiber and carbon black thermal paste, and culminating in my being the first person of Chinese descent to receive the Pettinos Award (an international award for research accomplishments in carbon science and technology), all indicate the superb and intricate design of my life journey by a higher power, namely my God, to whom I dedicated my life in 1976. I thank Him for guiding me, uplifting me, and, most of all, choosing me to be His instrument. I also thank my family (Fig. 1.14) for their prayers and support. In addition, I thank USA for the acceptance, education and research opportunities that I greatly value.

Chapter 2

James Chen Min Li: Leader in the Science of Engineering Materials

2.1 Introduction by the Editor

2.1.1 *What are engineering materials?*

Engineering materials refer to solid materials that are used in the making of a wide variety of products, including electronics, computers, sensors, data storage media, solar cells, aircraft, cars, roads, bridges, buildings, medical implants, batteries and machinery. Engineering materials constitute the foundation of technology, as appropriate choice and use of engineering materials are essential for attaining acceptable performance for any product. The history of human civilization, as marked by the stone, bronze and iron ages, shows that the advent of engineering materials brings about significant advances in human civilization. As technology progresses, new or improved materials that provide more demanding performance are needed. Even in the 21st century, engineering materials remain the controlling factor behind technological progress. In the aerospace industry, materials that are strong, stiff, lightweight, damage resistant and temperature resistant are needed. In the electronic industry, materials that are thermally conductive are needed for microelectronic cooling. In the biomedical industry, materials that are compatible with the blood and body fluids are needed for implants. In the solar energy industry, low-cost semiconductors in the form of thin films are needed. In the battery industry, electrodes that allow the reaction in the battery to occur reversibly are needed for rechargeable batteries. In the construction industry, concrete that can resist cycles of freezing and thawing is needed.

2.1.2 *What is the science of engineering materials?*

The development and use of engineering materials requires the making and modification of materials, the understanding of the microscopic structure within the material, the tailoring of the properties by controlling the structure, and characterization of the mechanical, electrical, magnetic, optical, thermal, chemical and other properties of the material. There are numerous types of materials, which include metals, polymers, ceramics, composites and semiconductors. The science of engineering materials encompasses physics, chemistry and engineering.

2.1.3 *Scientific contributions of Dr. Li*

The mechanical behavior of engineering materials are critical to the use of materials in structures, which include aircraft, cars, bicycles, ships, wheelchairs, turbine blades, machinery components, roads, bridges, buildings, etc. Mechanical behavior pertains to the strength, stiffness, toughness, fatigue resistance (i.e., resistance to repeated loading), scratch resistance, impact resistance, creep resistance (i.e., resistance to deformation at high temperatures) and vibration damping ability. The mechanical behavior of a material strongly depends on the internal microscopic structure of the material. Structural aspects include the crystal structure (i.e., the orderly pattern of arrangement of the atoms, ions or molecules in the solid), the grain structure (the size and shape of the grains, in case of a solid that consists of grains or crystallites), the imperfections, and the degree of crystallinity.

Dr. James Chen Min Li has made significant contributions to the understanding of how materials deform and become damaged under mechanical loading. Through extensive experimental and theoretical work, he has attained detailed understanding of how the structure relates to the mechanical behavior and how damage is initiated and evolves. His work covers numerous types of materials, including metals, ceramics, polymers and composites. The research of Dr. Li has resulted in about 347 publications in professional journals to date.

Dr. Li is outstanding not only in research, but also in education. He has been Albert A. Hopeman Professor of Engineering in University of Rochester (USA) since 1971. In his over 35 years of teaching, he has served as the major professor of 37 Ph.D. graduates to date.

2.1.4 *Honors received by Dr. Li*

Dr. Li is a member of the U.S. National Academy of Engineering. He received the Acta Metallurgica Gold Medal from ASM International (1990), the Lu Tse-Hon Medal from Chinese Society for Materials Science in Taiwan (1988), the Alexander von Humboldt Award for U.S. Senior Scientists from West Germany (1978), the Robert F. Mehl Gold Medal and Institute of Metals Lectureship from Metallurgical Society of American Institute of Mining, Metallurgical and Petroleum Engineers (1978), and the Champion H. Mathewson Gold Medal from American Institute of Mining, Metallurgical and Petroleum Engineers (1972). In addition, Dr. Li is a Fellow of American Physical Society, American Society for Metals, and The Metallurgical Society of American Institute of Mining, Metallurgical, and Petroleum Engineers.

2.1.5 *Career development of Dr. Li*

Dr. Li received his B.S. degree in Chemical Engineering from National Central University in Nanjing, China, in 1947. He went to work in Taiwan from 1947 to 1949. After that, he moved from China to USA, where he received his M.S. degree from University of Washington, Seattle, Washington, in 1951, and his Ph.D. degree from the same university in 1953. In 1953-55, Dr. Li was Postdoctoral Research Associate in University of California, Berkeley. In 1955-56, he was Supervisor of the MCA Research Project in Carnegie Institute of Technology. In 1956-57, he was Physical Chemist at Westinghouse Electric Corp. In 1957-69, he was Staff Scientist at United States Steel Corporation. In 1969-71, he was Manager of the Strength Physics Department, Materials Research Center, Allied Chemical Corporation.

Since 1971, he has been Hopeman Professor of Engineering at University of Rochester.

2.2 Dr. Li's Description of His Life Experiences

2.2.1 *Read the literature (Example 1)*

My first invention was at the Hsinying Paper Pulp Company in Taiwan in 1948 shortly after I graduated from college. I was assigned to work in a laboratory. This company made pulp from sugar cane bagasse, the left-over part after sugar cane was squeezed to get all the juice out for sugar manufacturing. Usually this left-over stuff is useless and only can be dried and burned as fuel. Then the Japanese found a way by cooking it in sulfurous acid to free the fibers which can be used to make paper. After cooking it was washed many times to remove the acid and shipped to paper manufacturing companies as pulp. In the cooking process, the acidity must be controlled very well. When it is too acidic, the fibers are weak. To change the acidity, soda and lime are added which produce carbon dioxide, a gas evolved with sulfur dioxide into the chimney. During those days the air pollution was not a concern but the workers often found that objectionable. Since the control of acidity was by hand, the product was not uniform. Sometimes the fibers were stronger and sometimes were weaker. Our laboratory was assigned to look into this problem.

Since I was not familiar with the pulp manufacturing process, I spent all my spare time reading in the library. One night I came across a Russian paper translated into Japanese. I do not read Japanese but I can recognize the Chinese characters and all the English chemical formulas. In this article it appeared to say that instead of acid you can use alkaline solution to make pulp. So the next day I tried sodium bisulfite instead sulfurous acid to cook the bagasse. It was an instant success. In the following weeks I and my technician cooked one batch after another with the vice president in charge of research, Mr. Chi-Ren Chou, watched daily because he was so excited. Within a couple months, it was tried on the manufacturing scale and the fibers produced were stronger than before. The workers were happy because they did not have to smell sulfur dioxide any more and they did not have to adjust the acidity. At that time there was no patent office in Taiwan so we did not have to apply for any patent.

A beneficial side effect was that I was famous over night in the company and caught the attention of a co-worker, Lily Yen-Cheung Wong, who was the "Company Flower" at the time and pursued by many eligible bachelors. I joined them and started dating her. I even got my first kiss not long after. Eventually I was the lucky one to marry her. Success sometimes breeds success. Your confidence and sincerity may win people. So this invention changed my life forever. Now you can see how important it is to read the literature.

2.2.2 *Read the literature (Example 2)*

In the summer of 1946 my mother was diagnosed to have diabetes. I was only 21 but my sisters and brother were even younger. So my father and I took up the responsibility of taking care of my mother. But my father's job required him to be away 6 days a week in the city of Chongqing, China and hence I was the one who took care of my mother most of the time. We made arrangements for a nurse who happened to be a neighbor to visit us daily to teach me how to make insulin injections and how to test the sugar in the urine. As an exchange, she brought her third grader son over so I could tutor him arithmetic. This way we did not have to pay each other.

Since I did not know much about diabetes, I started to go to a library and read. Gradually I knew it was due to the deficiency of the pancreas which could not supply enough insulin to digest sugar. In the western medicine, there was no cure. The patient had to watch her diet and receive insulin injections 3 times daily. Then one day I read in the Chinese literature about a secret formula for a cure. It involves eating raw pig pancreas daily. It sounds reasonable. So I decided to try it.

I made arrangements with a local butcher shop who killed a pig every morning at 7 a.m. I asked them to save the pancreas for me. To make sure the pig pancreas is fresh, I got up at 6 a.m. and put some wine in a covered bowl and carried to the chop shop and waited for them to kill the pig. I had to walk 20 minutes each way. After I came back I waited for my mother to wake up and made egg drop soup with the raw pancreas cut into small pieces and put in the soup just before my mother would eat it as breakfast. I did this every day for the whole summer. Amazingly I watched my mother's urine sugar improve while I reduced the insulin injection. By the end of summer, my mother's diabetes was practically cured. Her urine sugar was normal without insulin injections. I was so happy when I returned to college in the fall in Nanjing, China. In the meantime I heard that my student who entered 4th grade with his arithmetic grade improved from D to A. His mother was so appreciative and came over several times to thank my mother for that. So that was a productive summer for me.

While the story is true, I cannot certify that the secret formula worked because later in the year, my mother's diabetes returned. I was so far away I could not resume the treatment and there was nobody in the family who would have the confidence and patience to do that for her. So she returned to the insulin injection and the nurse came over again to teach my sister to do the injection and to test the urine. For the reader who has biomedical connections I would urge you to investigate the possibility of using pig pancreas to cure early diabetes.

2.2.3 Read the literature (Example 3)

At the National Central University (Nanjing, China), I took the thermodynamic course from Professor Chiang-Su Chang as an undergraduate. After I came to the United States, I took a graduate thermodynamic course from Professor Norman W. Gregory. Both were excellent teachers. So I consider myself well trained in thermodynamics.

In 1951, about 18 months after I arrived at the U. S. and enrolled at the University of Washington, I came across a paper about the temperature coefficient of the electromotive force (emf) of galvanic cells, i.e., how the voltage of a cell or battery change with temperature. I sensed something was missing, so I read some more and discovered that the author failed to see the general picture and made a conclusion with only limited application. I was very confident in myself, so I wrote a short paper in May 1951 and submitted to the Journal of Chemical Physics. But I was surprised that it was accepted right away and was published in August 1951, just 2 years after I arrived at the U. S. The chairman of the Chemistry Department at the University of Washington, Dr. Paul Cross, called me into his office one day and showed me the paper and congratulated me. He told me that I was the first graduate student in the history of the department who published a paper without the help of a professor. In fact, that was my very first paper published in the U.S. My Master's thesis was published in 1952 and my Doctoral thesis (a 4-paper series) was published in 1953.

That paper [1] stimulated some discussion in the literature for the next couple years so I had to publish another short paper in 1955 in the same journal [2]. My second paper was reviewed by several people and the editor had to consult Professor George Scatchard of Massachusetts Institute of Technology (M.I.T.) who assured him that I was correct. In the meantime, I had discussed with Professor K. S. Pitzer of the University of California, Berkeley, who agreed with me also.

2.2.4 *Take the challenge (Example 1)*

When I was a postdoctoral researcher at University of California, Berkeley, a graduate student, Pin Chang, came to see me one day in 1954 and presented to me a problem concerning the relation between viscosity and self-diffusivity in liquids. The relation was derived by Albert Einstein (American physicist, 1879-1955, Fig. 2.1) using a ball moving through the liquid. It agreed with the experimental result pretty well. Later, the same relation was derived by Henry Eyring (American chemist, 1880-1981, Fig. 2.2), based on his chemical kinetics (i.e., theory of the rate of a chemical process). The two relations differed by a factor of about 6. Both Einstein and Eyring were of course very famous scientists at the time. Even though I had a doctoral degree I was not much better than Mr. Chang. Can two young guys attack a problem not solved by two famous scientists? So I told him he was wasting his time. But he kept on coming back to me. He told me not to belittle ourselves too much. The problem might not be that hard.

Fig. 2.1 Albert Einstein   Fig. 2.2 Henry Eyring

Since I could not get rid of him, I started to look into the problem. I did not really know that much about viscosity even though I knew chemical kinetics and diffusion. I was doing low temperature heat capacity measurements with Professor Kenneth Pitzer at the time and I had two babies at home. So I did not really have that much spare time. But I studied and studied whenever I had a chance. One day it suddenly occurred to me that diffusion is a relative motion. An external force applied to an atom will make it jump ahead of its neighbors, whether the neighbors are stationary or moving. This little realization turned out to be exactly the factor of 6 that we were looking for.

I was so happy to tell Pin Chang the next day and we wrote a paper together in a few days and submitted to the Journal of Chemical Physics. It was accepted without any difficulty and published in 1955 [3]. A few years later, when Professor Eyring saw me at a meeting, he came over to thank me personally for solving this problem, which had bothered him for a long while. This paper was cited more than 100 times, as recorded by ISI Web of Science since 1973 and even cited in 2004.

I thank Pin Chang for the confidence in us and for taking up the challenge. Even the most famous still can overlook things, since they are too busy with other things or just have not had a chance to look into it. After all, a factor of 6 is not such a big deal, so it is just right for us two young guys to take the opportunity to make a name for ourselves. So trust yourself and take the challenge.

Fig. 2.3 Lord Kelvin   Fig. 2.4 Onsager

The problem surfaced again in 1996 almost 40 years later, when Fuqian Yang discovered that flow is much faster than diffusion under the same pressure gradient based on the current formulation. It leads to the proposal of a new Fick's law to solve self-diffusion problems [4].

2.2.5 *Take the challenge (Example 2)*

In 1956 I was at Westinghouse working on thermoelectricity. Irreversible thermodynamics was used to analyze the problem. Reading the literature, I discovered a conflict between the classical treatment such as that due to Lord Kelvin (Scottish physicist, 1824-1907, Fig. 2.3) and the modern treatment of Onsager (American chemist, 1903-1976, winner of the 1968 Nobel Prize in Chemistry, Fig. 2.4). The accepted analysis was that of Onsager and Kelvin's treatment was considered incorrect. Trained in classical thermodynamics I had high respect for Lord Kelvin. So I was puzzled and could not believe that Kelvin was wrong. Remembering the experience with Pin Chang a couple years ago, I took the challenge to resolve this conflict. I spent nights and nights reading and thinking. I remember I went to the basement after the kids went to bed and worked past midnight every day for all the summer. Finally I got it. Kelvin had a postulate in his mind which he considered natural but did not spell it out clearly. It turned out that this postulate is equivalent to Onsager's microscopic reversibility. I wrote several papers and continued working on that after I joined the E. C. Bain Laboratory where I had more freedom to do basic research. Eventually I discovered the thermokinetic potential [5] which decreases in all natural processes and becomes a minimum at the steady state, a parallel to free energy in isothermal systems which becomes a minimum at the equilibrium state. Later I used that concept in several microstructural processes in materials such as grain boundary migration [6].

2.2.6 *Take the challenge (Example 3)*

In about 1980 a dislocation-free zone at the crack tip was discovered in the electron microscope by S. M. Ohr's group at Oak Ridge National Laboratory (U.S.A.). This dislocation-free zone was inconsistent with the theory at the time. The theory was that of Bilby, Cottrell and Swinden, published in 1963, and was never questioned in the interim 17 years. Learning from my previous experience, I took the challenge again. The problem was difficult to deal with analytically, so we used a computer simulation. After some calculation, it was obvious that the difference between the classic theory and the recent experiment was in the critical stress intensity factor needed for dislocation emission. The classic theory assumed it was zero. By making it non-zero, a dislocation-free zone appears. The paper with Dai [7] has been cited 83 times since 1982. This realization stimulated a lot more calculations by computer simulation. We ended up publishing more than 30 papers in this area.

2.2.7 *Take a calculated risk*

When I was the manager of Strength Physics Department at the Allied Chemical Corporation Research Center, we were searching for new business ventures for Allied Chemical. In the summer of 1970, Dr. Ho-Sou Chen came to me looking for a job. Dr. Chen got his doctorate from Professor Turnbull at Harvard University and was doing postdoctoral research at Bell Telephone Laboratories. They did not have a staff opening at the time, so he could not stay. He was doing double roller quenching of liquid polymers. I recommended Dr. Chen to my boss, Dr.

Jack Gilman. But he was reluctant to hire him, since Dr. Chen's English was not that good. Later I talked to Dr. Chen further and tried to figure out something he could do. At the time, amorphous metal was a new material, but so far only small pieces could be made by splashing a liquid drop against a cold metal, or between two cold metal surfaces. While Dr. Chen was quenching liquid polymers, he could try to quench metals. To make it practically useful, he could quench liquid steels. With this idea in mind, I presented it again to Dr. Gilman, proposing only a summer job for Dr. Chen with the promise that he was going to make amorphous steel. While this was a definite possibility, I was not sure about it. If I did not make that promise, not only would Dr. Chen not get the summer job, but also the double roller experiment on molten metals would not have a chance to be tried. So I took a calculated risk in putting my reputation on the line to get Dr. Chen a summer job and a chance to try the process.

That summer, summer of 1970, marked the discovery of continuous casting of amorphous metals. The first batch, using carbon steels was a failure, the product was so brittle that it became powder easily. Then Dr. Chen added a lot of nickel. The product 2826 series was born. It was strong and yet ductile. Later, a soft magnetic property was discovered, which started a new business for Allied Chemical, namely, to make transformer cores.

After the summer Dr. Gilman offered Dr. Chen a permanent position, but after a year Dr. Chen returned to Bell Laboratories. By that time I was preparing to move to Rochester in the fall of 1971. However, my group at Allied modified the rapid quenching process for commercialization and that started a multi-million-dollar business for Allied.

2.2.8 *Benefit from group discussions*

Three shoe repair men are better than a general, as goes a Chinese saying. Sometimes a brain storming session or a group discussion may produce wonderful ideas. At the Edgar C. Bain Laboratory for Fundamental Research under the directorship of L. S. Darken, we were having a good time doing fundamental research without worrying about practical applications. One of the problems we were trying to understand was the equilibrium within stressed solids. The usual thermodynamic equilibrium deals with pressured solids or the stress is simply hydrostatic stress. What if the stressed state is not hydrostatic? Of course, we realized quickly that this was not complete equilibrium because plastic deformation would take place to relax the non-hydrostatic stresses. So plastic deformation was not allowed to take place. Hence this was a constrained equilibrium. An example was the concentration distribution of dissolved hydrogen, nitrogen or carbon in a bent piece of steel, or around a dislocation or a crack tip. However, we would like to formulate from macroscopic concepts, namely, thermodynamics without considering the atoms.

The three of us, Dr. Darken, Dr. Richard Oriani and myself, met frequently for a few months. Since Dr. Darken was the director and Dr. Oriani was the manager of the Physical Chemistry department in which I was a member, they were extremely busy. I was the only one with a more flexible schedule. But we met whenever and wherever we had a chance, at lunch, in my office after 5 pm, in Darken's office when we saw him for something else, some Saturdays or holidays when all of us were in the lab, etc. Dr. Oriani remembered a Moutier theorem and explained to us. All of us eventually made the critical distinction between mobile and immobile components and the rigorous definition of the chemical potential of mobile components. The problem was solved! Our paper was published in Z. Physkalish Chemie [8] in honor of Carl Wagner's 65th birthday. That paper was cited 384 times based on the Web of Science.

2.2.9 *Benefit from a group study*

Shortly after I joined the E. C. Bain Laboratory for Fundamental Research in 1957, dislocation theory was very hot at the time. Any paper in metallurgy or materials science was considered out of date if dislocation was not mentioned. Yet in the lab none of us knew anything about dislocations. So a few us decided to organize a reading group, each of us would present what he learned in the last period and we met every week. We had two books, one by Cottrell (1953) [9] and one by

Fig. 2.5 Dr. Li receiving the Acta Metallurgica Gold Medal in 1990 at the ASM International Fall Meeting in Detroit, MI (photo by Oscar & Associates, Inc.)

Read (1953) [10]. This went on for a few months. It turned out that I did most of the reading. But the idea of a reading group forced me to read regularly. It paid off because I wrote my first paper on dislocations (linear defects in a crystalline solid) in 1960. I made several presentations and published a series of papers. As a result I was invited to participate in the first international conference on "Electron Microscopy and Strength of Crystals" at University of California, Berkeley, in July 1961. My paper [11] on the "Theory of Strengthening by Dislocation Groupings", published in the proceedings of the conference in 1963, has had 157 citations based on the Web of Science. Imagine that I had not heard of "dislocations" before 1957. This invitation encouraged me to study further dislocation plasticity for the next few years. In 1967 I was invited to participate in the Battelle international conference on "Dislocation Dynamics." My paper on

"Kinetics and Dynamics in Dislocation Plasticity" [12] has had 328 citations based on Web of Science.

I must thank U. S. Steel Corporation, the Edger C. Bain Laboratory for Fundamental Research at U.S. Steel Research Center, and Drs. Richard Oriani and L. S. Darken for allowing us all the freedom to study. I built up my reputation entirely during the 12 years at U. S. Steel. It is regretful that this well known laboratory did not survive very long. Soon after President Kennedy allowed the union wages to increase but not the steel price, the profit of U. S. Steel fell and the lab was closed in 1971. I was fortunate to have joined Gilman at Allied Chemical in 1969, just 2 years before the ax. Dr. Darken was in tears when he announced the shut down.

Due to my reputation built up during the 12 years at the E. C. Bain Laboratory, I was invited to join the University of Rochester as an Albert A. Hopeman Chair in 1971 and was awarded the Acta Metallurgica Gold Medal (an international award for outstanding contributions in materials science) in 1990. The whole thing started with that timely group study.

2.2.10 *Interact with others*

Waiting to read the literature sometimes may be too late. That is why you should go to conferences to listen first hand the findings of others. The best is to interact with fellow workers in your own organization. This is not as easy as you think. Some people are reluctant to talk to you. So you must be generous to give ideas to others. If you give 10 and get 5 in return, you should be satisfied. Otherwise you must work alone and you would not get even the 5. So do not be afraid of giving away ideas. The more you can give, the more you would get in return. Some people may be as generous as you and you can publish jointly. Others may acknowledge your help or discussion. But some may just take your ideas and consider as his own. Do not worry about it. If you do not interact, you would never discover what kind of person he or she is.

An example was with Dr. William H. Hu of the E. C. Bain Laboratory. We had lunch together almost daily. One day he showed me a transmission electron microscope (TEM) picture of a sub-boundary (a boundary between slightly misoriented parts of a crystal), half of which almost disappeared. In the electron microscope, he could see a sub-boundary gradually faded away. This means that the two sub-grains on the two sides of the sub-boundary can merge into one grain. How could this occur? What is the driving force and what is the mechanism? I figured these out in a few days and this was how the "theory of sub-grain coalescence" was born. If I did not interact with him, I would miss the opportunity of analyzing this problem. He might have presented his findings at a conference, others may have analyzed it and the theory would not have originated from our laboratory. If I had remained ignorant and waited until he published his work, I would obviously have been too late to participate.

As a result of this interaction, his paper on "Annealing of Silicon Iron Single Crystals" [13] has had 253 citations and my paper on "Possibility of Subgrain Rotation during Recrystallization" [14] has had 179 citations since 1971 including 3 in 2004 according to the Web of Science. This problem has been rediscovered lately in the form of nanograin coalescence.

2.2.11 *Note of general concerns*

Dr. L. S. Darken used to tell me that the most important problem to solve is the one nobody else can solve and yet everyone wants to solve it. He himself did this with the Kirkendall effect and he is known for his famous "Darken Equation". See Thackray, 1986 [15] p. 201.

As I mentioned earlier, soon after I joined the E. C. Bain Laboratory for Fundamental Research, we were feeling the impact of dislocation theory. So we organized a reading group to study it. One of the concerns at the time was the source of dislocations. The Frank-Read source (a mechanism for generating dislocations in a crystalline solid) was generally accepted, but it was not generally observed. Instead, dislocations were frequently found near grain boundaries. Another concern was the yield stress-grain size relationship, generally known as the Hall-Petch relation, which was interpreted by using dislocation pile-ups (i.e., the pile-up of dislocations in front of an obstacle, such as a grain boundary). Yet Dr. A. S. Keh, a coworker at the Bain Lab. never observed any dislocation pile-ups in iron. But the Hall-Petch relation worked in iron and steels.

These were two major concerns in the field and, according to Darken's advice, I should work on it. So, one Saturday, I was thinking about these concerns and suddenly I found a way to connect them together, namely using grain boundary as dislocation sources. The Hall-Petch relation can be derived without dislocation pileups. Later I discussed with Dr. Keh who agreed with me. So I worked out the details of impurity and temperature effects and searched the literature for more experimental findings. They were surprisingly consistent. So I wrote it up.

The paper was published in 1963 [16] and was considered as a citation classic in 1986 (see Thackray, 1986 or Citation Classic, #49, December 7, 1981). It has been cited 283 times since 1972 including 3 in 2005 according to the Web of Science.

2.2.12 *Verify a famous equation*

A well accepted famous equation does not usually need verification, since, if you found it wrong, nobody will believe you. However, if you found it right, you can use that to substantiate your verification procedure. This is the motivation when I told my student, Indra Gupta at Columbia University, where I was a visiting professor in 1965-66 and an adjunct professor in 1966-71. I told him to analyze stress relaxation data using dislocation dynamics based on the Orowan equation.

Since I was still at the E. C. Bain Laboratory for Fundamental Research, Indra did his experiments in that Lab. However, before the paper was published in 1970, Indra got a job in Inland Steel and moved to East Chicago and I moved to Morristown, New Jersey, to join Dr. J. J. Gilman at Allied Chemical. We had some trouble to get the permission from the Bain Lab to publish under its sponsorship. (See Citation Classic, #39, September 23, 1989.) But Dr. Darken allowed us to publish without the U. S. Steel connection. So we published it under the Columbia and Allied Chemical sponsorships. We had no trouble at all with the reviewers and the key reader recommended publication without revision. See Gupta and Li [17].

The paper was cited 216 times, including one in 2004, according to the Web of Science, under I. Gupta. His results were included earlier in an invited paper presented at an international conference [18] which has been cited more than 400 times.

2.2.13 *Follow your intuition*

Your intuition is right sometimes and, if it is important to you, you should follow it. In 1963, during the peak of dislocation theory, I saw a paper with 10 pages of equations on angular dislocations. An angular dislocation is just a dislocation with two semi-infinite straight branches meeting at a point, so it looks like an angle. I looked at these equations with the intention to work out the energy and stresses of polygonal dislocation loops. But these 10 pages of equations scared me. However, when I looked at them more carefully, I had a hunch that they might be separable into the contributions of the two semi-infinite branches, because I saw some parts belonged to one branch and some belonged to the other. But it was not that clear due to the angle between them. So, one Saturday, I started to try for a simple case. It worked. For the next week, I tried one case after another. It all worked. Some were harder than others. But eventually I succeeded to separate all the results into the contributions of the two branches. The 10 pages of equations could be simplified to a couple of equations.

But there was a problem: a semi-infinite straight dislocation could not exist by definition. My stress field does not obey the equilibrium condition. So I worked out the body force needed for equilibrium. It turned out that this body force was the same for all semi-infinite dislocations of the same Burgers vector (vector describing the direction and magnitude of the distortion associated with a dislocation) that ended at the origin. So any two branches would cancel this body force. The problem was solved! I wrote to Elizabeth Yoffe at Cambridge University, who was the first working on angular dislocations. She did not believe me first, but later she checked with Eshelby, who assured her that my analysis was correct.

2.2.14 *Plan a group attack*

Sometimes, due to time constraints, you may not have the leisure to wait for new ideas to pop up. When I was invited to give the prestigious ASM weekend seminar in the fall of 1965, I had only about a year to prepare the lecture. Although I was at the E. C. Bain Laboratory for Fundamental Research and had many co-workers to assist me, I still had to perform some new experiments. So I applied for permission to hire two summer students in 1965, Mr. F. F. M. Lee to work on dipole annihilation and Mr. S. Sankaran to work on the annealing of dislocation density. It turned out that work from these gentlemen constituted 30% of my lecture [19]. Mr. F. F. M. Lee was so excited about his results that he continued and expanded to become his doctoral thesis after he returned to Stanford. His supervisor was William Nix at the time.

Another time, when I was invited to give the Institute of Metals Lecture (Robert F. Mehl award) in 1978, I also had only about a year to prepare. Fortunately I was at the University of Rochester, so I motivated all my students to help me. I had Sanboh Lee do analytical work and C. C. Chau do experimental work. I also had the help of Dr. S. N. George Chu, who analyzed some literature data for me. Their contributions were invaluable for the success of that lecture. This lecture [20], published in 1978, has had 121 citations, based on the Web of Science.

2.2.15 *Know your students*

You should know your student before you select a thesis topic for him or her. You should recognize his or her strengths and develop them further and also to strengthen his or her weaknesses. My philosophy is that a student should possess all of the necessary tools for research.   An example is Sanboh Lee, who was strong mathematically but weak experimentally. So he did a thesis analyzing diffusion-induced stresses and dislocation-crack interactions. However, I asked him to interact with Julie Harmon, who was doing experiments in the diffusion swelling of polymers and also to actually perform some experiments on mechanical healing of ionic crystals with Dr. Lothar Wagner, who was my postdoc at the time.   This combined experience provided him with a good background. So, when he went to teach at Tsing Hua University in Taiwan, he could supervise both theoretical and experimental research. As a result, he is now a Tsing Hua (Chair) Professor in his Department of Materials Science and Engineering. He was recognized in 1997 by the National Science Council in Taiwan as the best researcher and in 1998 was awarded the Roon Foundation award by the Federation of Societies for Coating Technology in the U.S. He was also awarded a Fellow of the ASM International in 2004.

Another example was Der-Ray Huang, who is a superb experimentalist, highly recommended to us by his superior at the Chung San Institute of Research in Taiwan. When he arrived in Rochester, he already had a family with two young kids. My funding on amorphous metals from the U.S. Department of Energy (DOE) was temporarily suspended, so I asked China Steel in Taiwan to support his research. They did and continued until he finished his Ph.D. degree. In this period, we applied for four patents in five countries on annealing techniques of amorphous metals to improve magnetic properties without affecting mechanical properties. However, I was not successful in finding him a suitable job in the U.S., so he went back to Taiwan. It was and remains a great loss for the U.S., because he made tremendous contributions in Taiwan and developed her optical storage disc industry. He has 77 patents to his credit and was awarded the Executive Yuan's Outstanding Sci.-Tech Talent Award in 1995, TECO Technology Award-Information Technology in 2000, Ten Outstanding Engineers Award by Chinese Institute of Engineers in 2001 and the Magnetic Technology Medal by the Taiwan Association for Magnetic Technologies in 2004. He is also a member of the Presidential Advisory Council for Science and Technology in Taiwan.

Another student of mine, Dr. Teh-Ming Kung, son of General and Mrs. Ling-Sheng Kung, did not return to Taiwan after his doctorate. His talent was recognized by Kodak at interview and he was hired on the spot. Now he is a Fellow at Kodak, the highest non-administrative rank. He has more than 50 patents to his credit and was awarded the Distinguished Inventor award in 1996. He is credited of developing the thermal printing process and all the related prize-winning products.

Kodak built several new factories based on his research, including one in Rochester.

All my students are productively engaged in teaching or research or both. I have had 6 from Columbia, 31 from Rochester and one from Nanjing University of Chemical Technology. In addition, I also have had 15 postdocs. An example is N. Balasubramanian, who obtained his Ph.D. degree from Columbia University in 1969. After working for a few years in the U. S., he returned to India in 1972 and started a program in advanced composites. This work was recognized in 1986 by a National Research and Development Council Award from the India government. He also developed non-asbestos technology and for that he was given the Nigel-Cutler award in 1985. In 1986 he published a book on advanced composite materials and in 1992, the Materials Research Society of India awarded him a medal for his work on composites. In 2002 he returned to his early interest in deformation mechanisms at Columbia and is now Emeritus Fellow of All India Council of Technical Education.

A few more examples: Mel Bernstein received his Ph.D. degree from Columbia University in 1965 and is now Director of University Programs—Science and Technology Directorate of the Department of Homeland Security. James B. C. Wu obtained his Ph.D. degree from University of Rochester in 1975 and is now Senior Vice President and Chief Technical Officer of Deloro Stellite. Sean Hsiang-Yung Yu received his Ph.D. degree from University of Rochester in 1976 and is now Director of Asian Research Office of the Army International Technology Center in the Far East. He was awarded the Fellow of ASM International in 2002. S. N. George Chu got his Ph.D. degree in 1977 and worked at both ATT Bell Labs and Agere. He had numerous awards for his achievements. In 1994 he was awarded the Fellow of the Electrochemical Society. Randy Wu Tung received his Ph.D. degree in 1979 and now owns Tung & Associates, LLC, specializing in intellectual property law. C. C. Chau received his Ph.D. degree also in 1979 and is now a Principal Scientist at Pactiv Corporation. Julie P. Harmon received her Ph.D. degree in 1983, worked at Eastman Kodak Company for 5 years and is now Professor of Chemistry at the University of South Florida. Paul Vianco received his Ph.D. degree in 1986 and joined

Sandia National Laboratories in the same year. He has published more than 120 papers and 6 book chapters, plus one Soldering Handbook. He is now a Principal Member at Sandia. Beta Yuhong Ni received her Ph.D degree in 1991 and is now program manager at Xerox, a training period for a vp position in the near future.

Fuqian Yang received his Ph.D. degree in 1995 and is now Associate Professor at the University of Kentucky. I had four students finish their Ph.D. degrees in 2001. Three of them are in Xerox Corporation; they are John Sheng-Liang Zhang and Jing Li (both Engineering Specialists) and Xinzhong Zhang (Mechanical Engineer/finite element analyst). The fourth one, Ping Zhu, is Materials Engineer at Delphi Corporation (a spin-off the General Motors Corporation). I was awarded the Graduate Teaching award from the University of Rochester based on the evaluation of my students.

Tong-Yi Zhang came to Rochester in 1988-90 and is now Professor of Mechanical Engineering at the Hong Kong University of Science and

Fig. 2.6 Dr. Li receiving the Graduate Teaching Award in 1993 at the University of Rochester

Technology. He was recognized as Fellow of ASM International in 2001 and as Fellow of the Hong Kong Institution of Engineers in 2003.

The successes of your students depend on their training in graduate school. By knowing your students, you can select a thesis and a training procedure to maximize their chances of success.

2.2.16 *Work hard*

You must work hard in order to achieve anything. You may be lucky sometimes, but even so, you must work hard in order to reach your full potential. In my elementary school years, I did not work very hard. My father gave me a book on crystal radios and the parts that came with it to make a radio. I loved that stuff and indulged in it any spare time I got, not just listening to the radio but also assembling it differently every time. We had a household helper, so I did not have to do any chores. The school work was easy too. I did not have to spend much time for it. Then the Japanese came and everything changed.

The crystal radio was not entirely a waste of time. When the government wanted us kids to learn how to prepare for Japanese air attacks, we all studied the booklet distributed to all citizens but since I had a radio I got a lot more information than others. So, when we competed in a city-wide written examination, I won the first prize. My school was very happy.

However, the realization that I did not work hard enough came in the summer of 1937, when I took the junior high school entrance examinations and did not get accepted by the two top Junior high schools in Soochow. So I realized that I was not as good as I thought. Then my father's bank, the Shanghai Commercial Bank, moved us in the fall of 1937 to Hupeh, in the city of Hankou. Suddenly we did not have any household helper. So part of household chores fell upon me, since my sisters were much younger and I had to take care of them too.

I remembered, during the summer months, since there was no school, I did all the cooking, cleaning, washing, grocery shopping, etc. every day, so my mother could have some rest, because in the fall when I went back to school, she had to do all these by herself. She was very appreciative that I could do these in the summer. The only free time I had was after everyone went to sleep. I utilized the couple of precious hours to study. It taught me a hard lesson not to waste any time. So when I went back to school in the fall, I studied very hard. I had to make up the time I lost in the summer.

From then on, I retained that habit. Even now I work six days in the office and the rest at home. Luckily I also have an understanding wife who takes care of all household chores. I help her clean house only on Sundays. During my school days I would bring geometry or other math homework with me, when I had to wait for anything. I would usually bring a paper or a book when I have to wait in an office. When a television show was boring, I would think on some recent problems I was encountering and shut my eyes. I frequently bring some problems in my mind when we go to a concert. Amazingly I have solved many problems while I was listening to the music, washing dishes, vacuuming the floor or walking. When the motion is monotonous and does not require brain function, the subconscious mind can work for you, if you store some problems in there.

A recent example was the grain boundary fluid flow model for the tin whisker growth. It bothered me for a few weeks, because it did not agree with the diffusion model already published in the literature. During one concert, it suddenly occurred to me that I did not have to solve the problem exactly. So, after I returned home, I tried an approximation. That was it, it agreed with the diffusion solution.

2.2.17 *Prepare yourself*

No matter how brilliant you are, you still need a proper environment to be productive. I realized this very early during my college years in China, since all the textbooks we used were written in English by foreign scientists. By the time I went to work in Taiwan in 1947, I decided I must go abroad. I was lucky to meet Lily Wong, who had similar inclinations. So we often talked about how to realize that dream. But there were no national examinations for overseas studies at the time. Many different schemes were mentioned but none was really practical. Then Lily's sister, Grace Wei-Cheung Wong, mentioned to me one day to write to important people and try my luck.

So I wrote to many government officers, including the governor of Taiwan and Madam Chiang Kai-Shek. I told them I had a new invention of cooking bagasse (see 2.2.1) and would like to improve the method further by going abroad. I got no reply except one, the secretary of the Governor of Taiwan, who invited me to visit the Governor in Taipei, Taiwan. But at the time I did not even have the travel money to go to Taipei from Hsinying, where I was working. Luckily, Mr. Po-Poh Sham, who had some savings and loaned me 500 Hong Kong dollars (about U.S. $100). With that money I went to Taipei. But the Governor was busy and had no time to see me. However, the secretary wanted me to stay in Taipei and wait. I stayed in a hotel and went to the Governor's office daily. After a week the secretary told me to return to Hsinying and wait. So I returned to Hsinying and expected that was the end of this plan.

To my surprise, the secretary wrote me again after about a month and told me to go to Taipei once more. I immediately got on the train and visited the Governor's office. The secretary told me that he had the Governor's authorization to support my round-trip travel to the U.S.A. and asked me to make an estimate. I gave a figure of a thousand U. S. dollars. He immediately gave me a slip of paper, which told the treasury of the Taiwan province to give me that money. I went to the treasury and the person who read the slip of paper did not believe it and had to call the secretary to make sure. But finally he gave the money to me. I came out of his office and could not believe what had just happened. I pinched myself to make sure that it was not a dream. I was so happy that, when I returned to Hsinying, I knew I was now a new person. In fact, I was so happy that I gave a friend of mine $50 because he had a large family and he was so poor that he desperately needed that kind of money. He said he would pay me back later. He did try to pay me back eight years later when I was already in the U. S. I told him to forget about it.

Mr. Po-Poh Sham has returned to Hong Kong, so on my way to visit my family, I stopped in Hong Kong and returned his 500 Hong Kong dollars. I must say that his help was critical. He took a risk because, if I was not successful, it might have taken years for me to pay back his money. My salary at the time was $10 a month and I could save at most

$2 a month. But he told me that he had confidence in me in view of my new invention.

My father was so impressed that he sold all my mother's jewelry to give me another $1,000, but asked me to pay him back in 2-3 years. I was so appreciative and was in tears when I accepted it. That was their life savings and they trusted me to pay them back, since my sisters and brother were not yet in college. What if I had trouble paying back this money?

I did pay back my father's money starting the second year I was in the U.S. Since he retired in 1950 soon after the communists took over China, I sent him monthly about $50 (less in the beginning and more in later years) until he died in 1997 at the age of 97. All my sisters and brother got their college education. I even helped their sons and daughters as much as I could. In fact, three of them are in the U.S.

For Taiwan, I returned many times to help her economic growth. I founded with Professor Lu Tse Hon the Chinese Materials Science Society in 1964. In fact I am still on the Advisory Board of their official journal "Materials Chemistry and Physics" published by Elsevier. In 1973 I helped create the Materials Research Laboratory in the Industrial Technology Research Institute. In the 1980's and 1990's many of my students went back to build up Taiwan's economy. Her $1000 dollar investment in me produced billions in actual benefits.

I told this story to demonstrate that, if there is a will, there is a way. My coming to the States was a miracle. But if I had stayed in Taiwan or in China, as many of my classmates did, my contribution to the society would have been much less. In fact, I may not have even survived during the Cultural Revolution in China. In the 50's and 60's I did help many of my classmates in Taiwan to come to the States.

2.2.18 *Get good grades*

Even though learning is more important than grades, good grades sometimes still help and in my case they were critical. After I arrived at the U.S., I entered University of Washington (UW), Seattle, in 1949. I had other admissions, including the University of Illinois. But I chose UW, because Lily Wong was in Vancouver, British Columbia, Canada, at the time. She had some Cantonese friends and actually emigrated to Canada. I studied very hard and earned straight A's. In some courses my final grades were very high compared to the second highest in my class. This impressed many professors, including Professor Ritter.

Lily Wong came over to Seattle in 1950 and we married in the August of the same year. A year latter our first son was born. Suddenly we were in financial trouble. Although my tuition was waived, our subsistence was only about $100 a month from the ECA (Economic Commission Administration, a U. S. government agency set up to help foreign students in the U. S. who could not return to their home country because of communists occupancy).

I went to several professors to look for help. Nobody could help me except Professor David M. Ritter, who personally took Lily to the employment office and then introduced her to a professor in the medical school, who gave her a job for $200 a month. Our problem was solved! Even though my interest was in physical chemistry and Professor Ritter's research was in inorganic chemistry, I decided to work for him for my Ph.D. thesis. I worked hard and wrote four papers for him, all published in the Journal of American Chemical Society, the best journal in chemistry. He granted my Ph.D. degree in 1953. I finished both master's and Ph.D. degrees in four years! He was very happy, because the student before me wrote only one paper with a wrong interpretation of the results. He used to tell others: "Jim Li has no limit; he can do anything." Actually of course I was just willing to learn. Shortly after I left, Professor Ritter was promoted to full professor. My four papers apparently contributed to his early promotion.

During my study with him, he helped us a lot in personal matters, including Lily's immigration problems. He wrote to the authorities and explained our situation. When we were both sick, he came over to deliver ginger ale. I remember that to this day. So, after he retired in 1987, I initiated a "Ritter Fellowship" in the Department of Chemistry at the University of Washington. It was set up so that he would designate a recipient (he was still active after retirement) until he died in 1997. But the "Ritter Fellowship" is permanent.

2.2.19 *Marry the right person and educate your children*

You may not realize how important is your spouse and family are to your productivity. I am lucky to have a loving wife who allows me to work hard on my career. Whatever my successes are, more than half the credit goes to her. I worked on Saturdays and most holidays. Many times I was the only one in the building. During the week days I was often late for dinner. At University of California, Berkeley, when I was a postdoc, once I had to do some experiments for a period of 36 hours non-stop. I remember Lily brought the kids with her to deliver meals to me. At the E. C. Bain laboratory when I had to finish a chapter or paper on time, I worked through midnight without going home for dinner. Lily again brought the kids and sent dinner to me.

I did spend one month at home after the birth of each child. Her complete recovery was important for her health. I also worried about my children's education and paid attention to their homework. All three of my children finished their Ph.D. degrees, one from Harvard, one from Columbia and one from Rochester. Their well being paved the way for

Fig. 2.7 Dr. Li and his wife, Lily Y.C. Li, taken in 1988.

their future. So do not neglect your family. Your productivity depends on them too.

2.2.20 *Get the proper rest*

My professor in Chinese literature at the National Central University told us often that those who work using their body should take a rest by using their mind. Those who work using their mind should take a rest using their body. So when I get tired of writing and reading, I work on my yard or clean the house and when I get tired of that I go back to my reading and writing. If you are a graduate student, when you get tired of sitting in front of a computer, go to the lab and do some experiments and, when you get tired of that go back to your computer. This alternating routine is especially good for combating old age. Both physical and mental activities are needed for the well being of both body and mind.

2.2.21 *Work across disciplines*

When the chemists started to work on metallurgical problems, they made important contributions in the thermodynamics of solid solutions. When people in mechanics and physics worked on mechanical properties of metals they made important contributions in dislocations and fracture theory. But it is not easy for a person to be versed in two or more disciplines. So the way to do that is to collaborate with someone in another discipline. However, you must learn enough to speak their language, so you would know what they are talking about and they would know what you are talking about. I started with chemical engineering in China, switched to physical chemistry at the University of Washington, learned low temperature calorimetry and statistical mechanics at Berkeley and studied irreversible thermodynamics at Westinghouse. After I joined the E. C. Bain Laboratory for Fundamental Research at the U. S. Steel Corporation, I studied dislocation theory and the elasticity which came with it, and learned physical and mechanical metallurgy. Typically you cannot solve a problem without the knowledge of several disciplines. So the more areas you know, the better is your chance of solving a problem.

2.2.22 *Summary*

I must say I have been very lucky in many of my ventures. I thank God for giving me the courage to pursue my achievements. I thank my wife for her patience, support and understanding in allowing me to work hard in my career. I thank my children for their successes and for editing this write-up. I thank all my professors, coworkers, friends and students for all their help and advice – too numerous to mention in this short memoir. For the reader, if I can do this much, you can do better. Just give it a try.

References

[1] James C. M. Li "Concerning the Temperature Coefficient of the emf of Reversible Galvanic Cells Operated at Variable Concentration" J. Chem. Phys. 19, 1059-1060 (1951)

[2] James C. M. Li "Concerning the Temperature Coefficient of the emf of Reversible Galvanic Cells Operated at Variable Concentration II" J. Chem. Phys. 23, 2012-2013 (1955)

[3] James C. M. Li and Pin Chang "Self Diffusion Coefficient and Viscosity in Liquids" J. Chem. Phys. 23, 518-520 (1955).

[4] Fuqian Yang and James C. M. Li "New Fick's Law for Self-diffusion in Liquids" J. Appl. Phys. 80, 6188-6191 (1996)

[5] James C. M. Li "Stable Steady State and the Thermokinetic Potential" J. Chem. Phys. 37, 1592-1595 (1962)

[6] J. C. M. Li "Irreversible Thermodynamics for the Motion of a Curved Grain Boundary" Trans. TMS/AIME 245, 1587-1590 (1969)

[7] Shu-Ho Dai and J. C. M. Li "Dislocation-free Zone at the Crack Tip" Scripta Met. 16, 183-188 (1982)

[8] J. C. M. Li, R. A. Oriani and L. S. Darken "The Thermodynamics of Stressed Solids" Z. Physik. Chem. Neue Folge 49, 271-290 (1966)

[9] A. H. Cottrell "Dislocations and Plastic Flow in Crystals" (Oxford: Clarendon Press, 1953) 223p

[10] W. T. Read, Jr. "Dislocations in Crystals" (McGraw-Hill, New York, 1953) 228p

[11] James C. M. Li "Theory of Strengthening by Dislocation Groupings" Chapter 15 in "Electron Microscopy and Strength of Crystals" (Ed. By G. Thomas and J. Washburn, Wiley, 1963) pp 713-779.

[12] J. C. M. Li "Kinetics and Dynamics in Dislocation Plasticity" in "Dislocation Dynamics" (Ed. By A. R. Rosenfield, G. T. Hahn, A. L. Bement, Jr. and R. I. Jaffee, McGraw-Hill, 1968) pp 87-116.

[13] Hsun Hu "Annealing of Silicon Iron Single Crystals" Proc. Symp. in New York, NY 1962, organized by the Physical Metallurgy Committee of The Metallurical Society of AIME, on "Recovery and Recrystallization of Metals" (Ed. By L. Himmel, Interscience, New York, 1963) pp. 311-378

[14] J. C. M. Li "Possibility of Subgrain Rotation During Recrystallization" J. Appl. Phys. 33, 2958-2965 (1962)

[15] A. Thackray "Contemporary Classics in Engineering and Applied Science" (ISI Press, Philadelphia, PA 1986) p. 220

[16] J. C. M. Li "Petch Relation and Grain Boundary Sources" Trans. Metall. Soc. AIME 277, 239-247 (1963)

[17] Gupta and J. C. M. Li "Stress Relaxation, Internal Stress and Work Hardening in Some Bcc Metals and Alloys" Metall. Trans. 1, 2323-2330 (1970)

[18] James C. M. Li "Dislocation Dynamics in Deformation and Recovery" Can. J. Physics 45, 493-509 (1967)

[19] James C. M. Li "Recovery Processes in Metals" chapter 2 in "Recrystallization, Grain Growth and Textures" ASM Seminar, October 16 and 17, 1965, American Society for Metals, Metals Park, OH, 1966, pp 45-98.

[20] J. C. M. Li "Physical Chemistry of Some Microstructural Phenomena", R. F. Mehl Lecture, Metall. Trans. 9A, 1353-1380 (1978)

Chapter 3

Paul Ching-Wu Chu: Inventor of High-Temperature Superconductor

3.1 Introduction by the Editor

3.1.1 *What is a superconductor?*

A superconductor is a solid material that conducts electricity without resistance, a phenomena known as superconductivity that was first discovered in 1911 by the Dutch physicist Heike Kamerlingh Onnes (1853-1926, winner of the Nobel Prize in Physics in 1913, Fig. 3.1). He found that the resistance of mercury becomes zero when it is chilled to a characteristic temperature (T_c) of four degrees above absolute zero, i.e. 4 K (-269°C or -452°F). As a result, a superconducting cable can be used for electricity transmission without energy loss over a long distance. This is in contrast with the conventional conductor that dissipates electrical energy and generates heat when electric current flows through it due to its resistance. A superconductor was later also found in 1933 by W. Meissner and R. Oschenfeld to expel an externally applied magnetic field, known as the Meissner effect, and thus to have the ability to repel a magnet. This repulsive force can be used for lifting objects, such as the levitation of a train above the ground. Superconductivity was recognized by F. London in the 40's and B. Josephson in 1964 to be a quantum phenomenon that enables the construction of ultra-sensitive detectors and ultra-fast switches from superconductors. Many applications that can profoundly impact our energy, medical, communication, transportation

Fig. 3.1 Heike Kamerlingh Onnes.

and defense industries have been proposed and demonstrated in the ensuing decades. During this period of time, many others metallic elements and intermetallic compounds have also been found to superconduct if cooled to below T_c.

3.1.2 *Invention of Dr. Chu*

In spite of the many applications of superconductivity proposed for decades, only a few have been commercially realized, such as the superconducting magnets used in magnetic resonance imaging (MRI) for medical diagnoses and high energy physics research, and the superconducting quantum interference devices (SQUIDs) used in magnetometers for the detection of minute magnetic signals for material research and medical diagnoses. This is because all known superconductors before 1986, regardless of their large number, have a T_c below 23.2 K (-249.8°C) that can be reached only with the help of the rare and costly liquid helium which boils at 4.2 K (-268.8°C) or marginally with the not-so-easy-to-handle liquid hydrogen with a boiling point of 20.3 K (-253.7°C or -422.9°F). The holy grail in superconductivity research in the seven decades following its discovery had been to raise the T_c to a practically high value, i.e. 77 K, a

Fig. 3.2 C.W. Chu (right) and M.K. Wu (left) in Chu's laboratory in University of Houston in 1979, while Wu was a graduate student.

temperature attainable by the use of liquid nitrogen which is plentiful, inexpensive and easy-to-handle.

In 1987, Dr. Chu and his co-workers (including Dr. Maw-Kuen Wu at the University of Alabama at Huntsville, abbreviated UAH, then, but presently at Academia Sinica Institute of Physics, Taiwan) invented the so-called "high-temperature superconductor" (HTS), which has a transition temperature T_c of $-180°C$. This "high temperature" is still much below room temperature (about 298 K or 25°C), but it is considerably above the extremely low temperatures mentioned above and can be attained by cooling, using liquid nitrogen (less expensive and easier to handle than liquid helium, with a boiling point of $-196°C$). The increased T_c greatly enhances the practicality of the use of superconductors. In addition, the invention forms a new class of superconductors, which are ceramics. In contrast, the conventional superconductors are mostly metals. Furthermore, the invention led to the creation of the modern field of high temperature superconductivity that has posed exciting challenges to condensed matter physicists and offered enormous technological promises to engineers. Fig. 3.2 shows a photo of Chu and Wu in Chu's laboratory in University of Houston in 1979.

3.1.3 *Contributions of Dr. Chu to science*

Dr. Chu's contributions to science go beyond the superconductor invention mentioned above and his research interest spans from basic to applied. He has made numerous contributions to Experimental Solid State Physics, particularly in superconductivity, magnetism and dielectrics. (Dielectrics refer to electrical insulators.) His studies addressed the structure and behavior of many compounds and alloys, and made use of a large variety of experimental techniques, especially at low temperatures under high pressures. He has succeeded in unraveling the physics of many novel materials, discovered new phases and synthesized new compounds under pressures that do not form otherwise. Dr. Chu has established himself as a world leader in the areas of solids at low temperature and high temperature superconductivity. He is also engaged in bringing the exciting discoveries in the laboratory to practical applications to benefit the society, e.g. he is heavily involved in the deployment of the high temperature superconducting cable for electrical power transmission. He has over 560 refereed articles published in scientific journals.

3.1.4 *Contributions of Dr. Chu to education and research organization*

Dr. Chu has served the education and scientific communities in various capacities. For example, he is the President of Hong Kong University of Science and Technology (HKUST) (2001-present). He served as the Convenor of the Heads of Universities Committee in Hong Kong (2003-4), Member of the Steering Committee on Innovation and Technology Commission of Hong Kong Government (2003-present) and member of the Board of Hong Kong Science Park (2002-5). In addition, Dr. Chu holds the T.L.L. Temple Chair of Science at the University of Houston (1987-present) and is the Executive Director of the Texas Center for Superconductivity at the University of Houston (2005- present) that he founded in 1987 and served as its Director from 1987 until taking up his presidency in Hong Kong. He was Director of the U.S. National Science Foundation Materials Research Science and Engineering Center on Advanced Oxides and Related Materials at the University of Houston

(1996-7), Director of The Texas Center for Superconductivity at the University of Houston (1987-2001), Member of the Texas Governor's Science and Technology Council (1994-2000), Director of the Solid State Physics Program of the U.S. National Science Foundation (1986-7), Member of Board of Directors of the the Committee on Superconductivity for America's Competitiveness (1988-present) and Director of the NASA Space Vacuum Epitaxy Center at the University of Houston (1986-89). He also serves on various committees and advisory boards of institutes, professional organizations and journals in the U.S. and abroad.

3.1.5 *Honors received by Dr. Chu*

Dr. Chu received numerous awards and honors in recognition of his contributions to science. They include the U.S. National Medal of Science from President Ronald Regan in 1988 (Fig. 3.3), NASA Achievement Award in 1987, Comstock Award from U.S. National Academy of Sciences in 1988, International Prize for New Materials from American Physical Society in 1988, Medal of Scientific Merit from World Cultural Council in 1989, Texas Instruments Founders' Prize in 1990, Best Researcher in the U.S. from US News and World Report in1990, St. Martin de Porres Award in 1990, Superconductivity Award of Excellence in Scientific Accomplishments from World Congress on Superconductivity in 1994, Houston Hall of Fame Award in 1999, John Fritz Medal from American Association of Engineering Societies in 2001, etc. In addition, Dr. Chu has been elected as a Foreign Member of Chinese Academy of Sciences, a Member of the U.S. National Academy of Sciences, a Member of the American Academy of Arts and Sciences, a Member of the Russian Academy of Engineering, a Member of Academia Sinica (Taiwan), and a Fellow of American Physical Society. He received honorary doctorates and professorships from many universities in USA and China.

To Paul Chu
Congratulations and best wishes,
Ronald Reagan

Fig. 3.3 Dr. Chu received the U.S. National Medal of Science from President Ronald Regan in 1988.

3.1.6 *Career development of Dr. Chu*

Dr. Chu was born in Hunan, China, in 1941. His parents were from the Guangdong Province. He received B.S. degree from National Cheng-Kung University, Taiwan, in 1962, M.S. degree from Fordham University, Bronx, NY, USA, in 1965, and Ph.D. degree from University of California, San Diego, CA, USA, in 1968. All of his three degrees are in Physics.

After obtaining his Ph.D. degree, Dr. Chu worked as a member of the technical staff of Bell Laboratories in Murray Hill, NJ, USA (1968-70). Then he became Assistant Professor of Physics in Cleveland State University in Ohio, USA (1970), followed by promotions to Associate Professor (1973) and Professor (1975). In 1979, he became Professor of

Physics at University of Houston (1979-present). Dr. Chu also conducted research in Argonne National Laboratory, Argonne, IL, USA (1972), Stanford University (1973), Los Alamos Scientific Laboratory, Los Alamos, NM, USA (1975-80), University of California, Berkeley, CA, USA (1991), and Lawrence Berkeley National Laboratory, California, USA (1999-present).

Dr. Chu is married with two children. He speaks English, in addition to Cantonese, Mandarin and a few other Chinese dialects.

3.2 Dr. Chu's Description of His Life Experience

3.2.1 *Happy boyhood in Ching-Shui*

My parents moved the family to Taiwan in 1949 to escape the chaos in the Chinese Mainland after the Nationalist Chinese Government was driven out from the Mainland. We settled in the small sleepy town called Ching-Shui on the west coast in central Taiwan. The town acquired its name because of the quality of its water since Ching-Shui means "clear water". I cannot believe that, by fate or coincidence, fifty-two years later I came to the Hong Kong University of Science and Technology which is located by the beautiful "Clear Water" Bay. In 1946 Winston Churchill introduced the phrase of "Iron Curtain" to describe the division between the Western powers and the area controlled by the Soviet Union at Westminister College, in Fulton, Missouri after receiving his honorary degree. Cold war started with the rapid expansion of the Soviet influence into Eastern and Central Europe afterward. The communist's victory in China in 1949 and the release of the white-paper by the U. S. government to show its no-confidence in the Nationalist government created a sense of abandonment by the world for the people in Taiwan. Not until 1950 when Korea War started and the U. S. sent her seventh fleet to Taiwan Strait, such a feeling started to abate. In spite of this global political earthquake, Ching-Shui provided the much needed peace and tranquility for me to grow happily till I left for the U.S. for graduate study in 1963 after a year of military service as a second lieutenant in the

Fig. 3.4 Chen Ning Yang   Fig. 3.5 Tsung-Dao Lee

air force as required for all college graduates at the time. The town had the tradition of more on traditional mandarin pursuit than on commerce. I grew up in an environment that was spiritually rich although materially poor. I spent many weekends and holidays to go bird-hunting, fishing and swimming in the forest, rivers and the Pacific Ocean after study. It helped me develop the life-long interest in plants, flowers and gardening. I graduated from the local primary school and high school and went south to attend the National Cheng-Kung University in the old city of Tainan.

It was rather natural for my generation to study science or engineering after facing the repeated disastrous defeats of China by the technologically-superior West in late 19th and the first half of the 20th century. We wanted to help build a prosperous and strong China. I have always had a keen interest in things involving electricity and magnetism, since I could remember. However, it was the award of the 1957 Nobel Prize to two young Chinese physicists, Chen Ning Yang (Fig. 3.4) and Tsung-Dao Lee (Fig. 3.5), that made my career choice between an electrical engineer and a physicist easy.

The influence of my technically oriented father, Mr. Kan-Ting Chu, could not be overstated (Fig. 3.6). He was trained as a pilot in the U.S. and returned to China to join the Nationalist Air Force to fight the Japanese invasion in the mid-1930's.

Before attending kindergarten, I already watched him with fascination to repair electrical appliances and mechanical gadgets at home, often to mid-night. I started making crystal radios when I was

Fig. 3.6 Mr. Kan-Ting Chu, father of Dr. Chu, in 1928

in primary school and motors in junior high school, going to the junk yard to pick up the materials and parts needed. I still remember the exciting moments when I could listen to the banded broadcasting from The Central People's Radio Station in Beijing into the wee hours in the morning after everyone in the family went to sleep, using my primitive radio. Equally exciting was when the motors turned after connected to the battery (for the dc one) or plugged to the power socket (for the ac one). I could just sit by the motor for hours tirelessly listening to the hum of the rotating motor. I even dreamed of building a machine that would move forever once started even though books from the library and teachers all said it was an impossible dream. I had been fascinated by science fiction writings based on such machines. I still remember the countless nights in pursuit of such a dream. I connected in parallel the two identical motors that I built and prayed that once one was rotated as the generator, it would produce enough power to keep the other rotating without stopping. Clearly, I failed but never was I despaired. The experience has something to do with my later decision to become an experimental physicist with an empirical inclination. This was in contrast to the traditional Chinese mandarin thinking, i.e., those using their brains were considered to be superior to those using their hands.

3.2.2 *Great teachers in Taiwan*

Chinese has a saying "one has to be at the right moment, in the right place with the right people" to be successful. I believe that I have been extremely fortunate to have many great and dedicated teachers to learn from, and colleagues and friends to work with in the right places and at the right time over the years. A few examples that follow may reflect the great fortune I have had.

During the last two years in the Ching-Shui public primary school, Mr. Min-Tser Hsu was our class-teacher. He graduated from the First Tai-Chung High School, the best high school in central Taiwan, near the end of the Second World War. His dream to become a scientist by going to college in Japan was crushed due to the heavy Allied bombing on the sea lane between Taiwan and Japan. It appeared that he wanted to realize his dream in his students. Educated in the Japanese system, Mr. Hsu was a strict disciplinarian. He taught us the three-step approach to problem solving, i.e., to understand the problem, to solve the problem via as many routes as possible and to check and contrast answers from the multi-routes before picking the best and elegant answer. To date, I am still benefited from this approach.

I was equally lucky to have inspiring and caring teachers in the six years in Ching-Shui High School like Mrs. C. J. Huang, D. F. Jin, T. C. Liu, Y. J. Tsai and B. H. Wu, who taught us to think outside the box and at the same time to be a good citizen. They encouraged us to examine problems outside the textbooks and worked with us when we hit the wall. I still remember the thrill I had when these problems were solved, especially by myself. To do and achieve what I am not required to do or to obtain results others do not expect always give me extra satisfaction and has been what I try to practice throughout my life.

Before the late 70's, there were very few PhD's in Taiwan, even fewer in universities, and research was almost nonexistent, due to dire economic condition in Taiwan. However, university professors still were considered to be the elite and command great respect from the society.

They were extremely enthusiastic and dedicated teachers. Those that come to my memory include Professors T. S. Chang, W. C. Pang, S. Y. Wang and K. D. Yang who had provided us a rather solid physics background and shared with us the excitements of physics although they themselves were mere observers. I was fascinated by the semiconductor radios and decided to study the electrical properties of pure germanium when I took the solid state physics class with Professor W. C. Pang in my junior year. To get the crystal, we had to order it from Japan. Pang went through the enormous bureaucratic mess to get US$3,000 by overcoming the strict government foreign currency control, not to mention that the sum was astronomical in comparison with the monthly pay of the professors. It took almost two years to receive the crystal when I was about to graduate. Needless to say, the experiment was never done. However, Pang's effort to help and inspire a student has impressed me immensely and has served as a good example for me later to be a teacher.

3.2.3 *Great teachers in the U.S.*

The graduate study in the U.S. forever changed my life. In 1963, I received several graduate assistantships to study in the U. S. – the dream for the great majority of college graduates in Taiwan those days. I chose Fordham University to study physics because my friend, John Wong, went there a year earlier and a Nobel Laureate, Professor V. F. Hess, was there. The physics faculty was young (except Hess), active and friendly, in contrast to my prior perception of scientists as being like Einstein. Among others, Professors Joe Budnick and Joe Mulligan had had a decisive role in shaping my future physics career. Budnick was young and energetic and had a broad knowledge in solid state physics. I joined his group to examine the magnetic hyperfine structures of solids and had a chance to meet and learn from many prominent colleagues of his. The excitement of research in his laboratory attracted me to the field of solid state physics. After a year at Fordham, I learned that the three most prominent physics departments in solid state physics in the U. S. at the time were located in University of Illinois at Champaign-Urbana, University of California at San Diego (UCSD) and Brown University. Thinking about the possibility to learn directly from those whose names I

Fig. 3.7 B.T. Matthias in La Jolla, CA, in 1980.

Fig. 3.8 J.G. Bednorz   Fig. 3.9 K.A. Müller

could find in the textbooks (my reference as great physicists), I went to Professor Joseph Mulligan, the Department Chairman, and asked for advice. He   happened to have spent a wonderful and productive sabbatical year at UCSD in 1962 and strongly recommended that I should consider seriously furthering my graduate study there. Years later in the mid-90's, I saw Professor Mulligan in a school reunion in New York. He told me he found the recommendation letter several weeks earlier when he cleaned up his office and was proud of what he wrote for me then, although he was torn at the time – to keep a good student for the department or to let go the student for his better future. Clearly he had in his mind the student's welfare ahead of the department's.   So did Budnick. I was moved.

Solid state physics (or condensed matter physics) has been my favorite subject by design since my college days in Taiwan because of its dual nature: its intellectual challenge and technological promise. However, superconductivity became my lifetime "hobby" by accident, because of the late Professor Bernd T. Matthias (1918-1980, Fig. 3.7), after I entered the University of California at San Diego as a graduate student in the summer of 1965. At that time, choosing superconductivity as the thesis topic was much harder than it seems to be now. In the '60s, to many young graduate students, "physics" meant nothing more than particle physics, which dealt only with the most profound and most fundamental unknowns of the universe.   Charles Kittel wrote in his *Introduction to Solid State Physics* that almost all forces in action in a solid are basically electromagnetic and known.   The study of solids just did not seem to be able to arouse excitement in the young inquisitive mind of a graduate student on the same level as an examination of elementary particles.   Only later, in 1966, did I find out how similar the paths were of Murray Gell-Mann, who predicted elementary particles based on his eight-fold way, and Bernd Matthias, who searched for high temperature superconductors based on his empirical rule.   I finally came to realize that no subject is too insignificant to work on, and that the significance of the subject depends only on how it is worked on.

Until his untimely death in 1980, Matthias was an undisputed leader in the field of superconductivity. His remarkable insight into the physics and chemistry of materials was the only effective guide in the search for superconductors with a higher T_c before the seminal work of J. G. Bednorz (1950- , Fig. 3.8) and K. A. Müller (1927- , Fig. 3.9) in 1986 (winners of the1987 Nobel Prize in Physics). His unceasing enthusiasm and optimism about the future of superconductivity provided much needed adrenaline for the field during the dog days of the late '70s and '80s, when funding diminished to a trickle. He was a strict Edisonian and was deeply skeptical about theories.   As a former student of his, I havecontinued to be greatly influenced, for better or for worse, by his style of doing physics and his taste in selecting problems.   Throughout my career in condensed matter physics in general and superconductivity in particular, I have benefited significantly by paying attention   to materials which are considered too mundane by many traditional physicists, and by not being intimidated by theoretical predictions which are often treated as Sacred Writ by many experimentalists.   I feel extremely blessed to be in the right place at the right time so that I could witness one of the most exciting developments in physics and even played a role in it.   For that I am forever grateful to my mentor, the late Bernd Matthias, and many dedicated and hardworking colleagues in Houston.

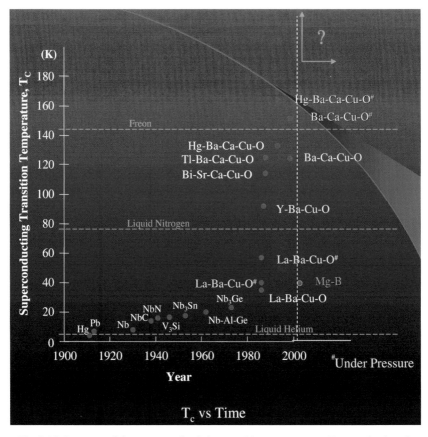

Fig. 3.10 Increase of the superconductivity transition temperature T_c over the decades.

3.2.4 *The torturous path to high temperature superconductivity (HTS)*

To seek materials with higher T_c, to unravel the physics of superconductors, and to develop superconducting devices define all activities in the field. The study of superconductivity over the years has continued to generate excitement in various areas of condensed matter physics, ranging from the discovery of novel compounds, through the detection of new physical phenomena, to the development of new theories. However, the road to compounds with a high T_c was torturous before 1986 (Fig. 3.10). The record T_c in 1986 remained 23 K (-250°C) which was set in 1973, representing only a 19 K (19°C) increase after a continual effort over three quarters of a century. However, the situation was drastically changed after 1986.

3.2.4.1 *Before 1986*

A record T_c of 23 K was significant at the time because it put us over the threshold of liquid hydrogen cooling. Superconductors with such a relatively high T_c then were all intermetallic. They are now known as the conventional low temperature superconductors (LTSs) which distinguish themselves from the non-intermetallic high temperature superconductors (HTSs) discovered after 1986 with a much higher T_c. Prior to 1986, there were two general approaches to following the search for superconductors with a higher T_c: the BCS approach and the enlightened empirical approach.

The BCS approach
It follows the first microscopic theory developed in 1957 by John Bardeen (1908-1991), Leon Cooper (1930-) and Robert Schrieffer (1931-) (BCS) (winners of the Nobel Prize in Physics in 1972, Fig. 3.11) on superconductivity, known as the BCS theory. The theory provides the microscopic mechanism (electron-phonon interaction) to explain the origin of superconductivity and accounts for many of the superconducting properties of the LTSs, but fails to show where and how to find a high T_c superconductor. The theory does predict that a strong electron-phonon interaction can lead to a high T_c, but too strong such an interaction will result in a crystal structural transformation. A maximum

Fig. 3.11 John Bardeen (left), Leon Cooper (middle), and Robert Schrieffer (right).

T_c not exceeding the 30's K (the -240's °C), due to the crystal structural transformation was therefore proposed. While the proposition appeared to be reasonable based on the observation of structural transformation in the then high-T_c intermetallic superconductors, no direct evidence was available. We decided to control and even remove the structural transformation, using the high pressure technique, and to examine directly its effect on T_c. To our great relief, we found in the late 1970's that the effect was small and not more than a few tenths of a degree in the then HTSs, V_3Si and Nb_3Sn with T_c's of 17 K and 18 K, respectively. The observation provides the intellectual basis for my belief in the possibility of a higher T_c and the effectiveness of employing the high pressure technique to make such a possibility a reality. It happened to be the high pressure results on the T_c of the first HTS that demonstrated the importance of cuprate in high T_c and that provided the rationale for chemical substitution to achieve higher T_c.

In the 1970's and 80's, various nonconventional mechanisms and possible material systems were proposed for higher T_c. They included low dimensional, multilayer, unstable, and oxide material systems. I had studied extensively on many of these material systems but did not find a higher T_c. Despite the failure to raise T_c, the decade-long persistent hard work described above did give my group a proper perspective on the high T_c problem, skills in synthesis and characterization, and knowledge of the physics and chemistry of materials in general and of oxides in particular. This experience had helped immensely in our later discoveries in HTS's.

The empirical approach

The so called enlightened empirical approach has served us all in the field well in our long search for materials with higher T_c. The approach depends on the insight and knowledge of the practitioner. The most successful example of the enlightened empirical approach before 1986 was due to Bernd Matthias, the so-called "Matthias empirical rule" proposed in 1953. The rule correlates T_c with the ratio of valence electrons to number of atoms (e/a). According to the rule, the maximum T_C of a compound system unusually occurs at e/a-ratios of 4.75 and 6.4. Hundreds of intermetallic compounds and alloys were accordingly discovered with a T_c up to the then record value of 23 K. For instance, Nb_3Sn (T_c = 18 K) and Nb_3Ge (T_c = 23 K), all have a e/a-ratio of 4.75. No T_c-(e/a) correlation similar to the Matthias empirical rule has been found among the HTSs. However, the general enlightened empirical approach, sometimes aided by theoretical speculations, has worked wonders later in the field of HTS, as mentioned in the paper of Bednorz and Müller in 1986, although Matthias empirical rule does not.

3.2.4.2 *The critical year of 1986*

Disappointed by the slow progress in raising T_c over several decades, Müller and Bednorz decided to deviate from the conventional intermetallic path and dive into the oxides in an attempt to take advantage of the enhanced electron-phonon interaction in these compounds. In January 1986, they detected superconductivity up to 35 K (-238°C) in their multiphase La-Ba-Cu-O (LBCO) samples, setting a new T_c-record of 35 K. The superconductivity was later attributed to the $La_{2-x}Ba_xCuO_4$ (La-214) phase. The appearance of superconductivity at such a high temperature in an oxide surprised many experts. The work opened a new chapter in high temperature superconductivity (HTS).

The report did not initially attract the attention it deserved, in part because of the general belief of unlikelihood of high T_c in oxides, the many false alarms of high T_c previously reported and their modest title, "Possible High T_c in the La-Ba-Cu-O System". However, I read the paper with immense excitement on November 6 and felt that this was the dream I had pursued over the preceding 10 years, to set a T_c-record in non-intermetallic compounds. It did not take much effort for my group to reproduce and confirm the results in $La_{5-x}Ba_xCu_5O_x$ (LBCO) with Bednorz and Müller's compositions, on November 21. It is hard to believe that the greatest experimental challenge at that time was the thermometers, which were all valid only below 25 K. We had to calibrate one of the Ge-thermometers against a copper-constant thermocouple and a chromel-alumel thermocouple to be sure that the temperature was correct. An even more exciting observation was made only four days later on November 25 in another multiphase sample of LBCO, i.e. a large resistance drop by a factor of 80 at ~ 75 K (~ -198°C), suggestive of a superconducting transition associated with an unidentified impurity phase that was different from the 30 K (-243°C) La-214 superconducting phase. By examining the pressure effect on the T_c of the 30 K (-243°C) samples, we found an unusually large pressure effect, more than ten times that detected in previously known superconductors, and set a new record T_c of 40.2 K and later to 52.4 K by pressures. The observations showed that these superconductors are different from the conventional ones and the new record T_c exceeds the one predicted by the BCS theory. We decided to search for higher T_c in these and related oxides.

In early December, at the 1986 Fall Meeting of the Materials Research Society Meeting, I showed Maw-Kuen Wu, my former student and then an assistant professor at the University of Alabama at Huntsville (UAH), our data and invited him to join me in the search for even higher T_c. This started a very fruitful collaboration.

I dropped Alex Müller a note after duplicating their results in early December. About a week later, Alex called me from Switzerland after receiving my note. He thanked me for confirming their data. I thanked him for providing a new direction for us to bring the T_c to a new height. I also told him about our high pressure results and that 77 K was a strong possibility without telling him explicitly about our observation of occasional resistance drops above 60 K. We chatted briefly about the important role of faith in the discovery of superconductivity at ever higher temperatures. Later, in March 1987, Alex confessed to me that he thought, at the time, that I was over-optimistic.

3.2.4.3 *The exciting 1987 & the "Woodstock of Physics"*

As soon as we detected an unusually large pressure effect on the T_c of LBCO and observed sign of superconductivity up to 75 K in the multiphase LBCO samples, I thought of simulating the pressure effect by chemical means to retain and stabilize the high T_c, i.e., replacing Ba and La in LBCO, respectively, by the isovalent but smaller Sr and Ca, and non-magnetic rare-earth elements Y, Yb and Lu. Substituting Ba with the smaller Sr was immediately shown to raise the T_c by us and later by others. Prodding by the University, a patent application based on the above idea was filed on January 12, 1987. My faith in higher T_c was boosted on the same day by the detection of a rather strong Meissner signal in one of our multiphase LBCO samples above 90 K. This was the first unambiguous superconductivity signal above the liquid nitrogen boiling point of 77 K ever observed. Unfortunately, the signal disappeared the next day, due to the high impurity of the sample. We immediately turned our attention to stabilizing the 90 K phase by chemical substitution. We did take the X-ray diffraction pattern of the sample (identified as the La123 phase later) but failed to determine the crystal structure at the time.

While we were waiting for the rare-earth elements, I started writing a manuscript to report our 90 K superconductivity observation in the LBCO samples. I thought that it should be fine to publish the preliminary data so others with better ideas and equipment could stabilize the 90 K superconducting phase, provided we stated clearly the conditions under which the phenomenon was observed. We received an exciting call from Maw-Kuen (Fig. 3.3) from UAH at about 5 p.m., January 29, 1987. He informed us that he and his students had just observed a reversible sharp drop in resistance, starting at above 90 K and finishing at about 77 K in two of their samples. All of us were ecstatic, since stable and reversible superconductivity might finally be realized, if Meissner effect could be detected. Maw-Kuen and his graduate student Jim Ashburn arrived at Houston the next morning on January 30 with their multiphase samples which had a nominal formula of $Y_{1.2}Ba_{0.8}CuO_4$ (abbreviated YBCO). We subjected them to a thorough battery of tests. Indeed, the resistance decreased rapidly, starting at ~ 93 K and reached zero at ~ 80 K, as was first observed at Huntsville by Maw-Kuen and his students. By the end of the day, we completed the magnetic measurements. A distinct diamagnetic shift characteristic of a superconducting transition started at ~ 90 K. Several samples were also made in Houston and tested on January, showing the same 90 K superconducting transition. The long-sought-after stable and reproducible superconductivity above the liquid nitrogen boiling point of 77 K was discovered. Peiherng Hor, my student, determined the pressure effect on the T_c of YBCO samples and found that it was very small, which was drastically different from the La-214 phase. The X-ray diffraction pattern obtained by Ruling Meng, my Research Associate, was clearly different from the 30 K La-214 phase. We concluded that the 93 K superconductor YBCO must be a new phase. The results were submitted on February 5 and appeared in the March 6 issue of Physical Review Letters.

The next challenge was to isolate the 90 K phase and resolve its structure, since the samples consisted of many phases. After failing to resolve the structure ourselves, I requested help from Dave Mao and Bob Hazen at the Geophysical Lab, who had extensive experience solving structures of tiny crystals. Mao and Hazen crushed the multiphase samples, singled out the tiny crystals and characterized them. At Houston, Peiherng and Ruling worked day and night with students to prepare samples with greater volume fractions of the superconducting phase. By comparing the data on X-ray and the size of the superconducting signal of samples with various compositions of Y:Ba:Cu under the same synthesis condition, the superconducting phase was identified to have a stoichiometry close to 1:2:3 in late February. On March 1, Hazen et al. finally determined the exact stoichiometry and the structure with all the cation sites assigned except oxygen, showing a triple layered structure ($YBa_2Cu_3O_{7-\delta}$ or Y123). Many other groups worldwide obtained the same structure with the oxygen sited also determined at about the same time or later.

Once the structure was determined, we decided to probe the role of Y by following the conventional way, i.e., to replace some Y with the magnetic rare earth elements. We found that, even with a large fraction of Y replaced by the magnetic Gd and Eu, no T_c depression at all was detected in contrast to expectation for LTSs. This suggested that Y in Y123 is electronically isolated from the superconducting electron system and serves mostly as the stabilizer in the compound crystal. It was natural for us to attempt to synthesize $RBa_2Cu_3O_{7-\delta}$ (abbreviated R123) with R = rare earth elements. We succeeded in our first attempt to obtain a new series of 90 K superconductors R-123 with R = Y, La, Nd, Sm, Eu, Gd, Ho, Er, and Lu. According to the results, we recognized the significant role for the CuO_2-layers in the 90 K superconductivity in R-123. Re-examination of the X-ray pattern of the LBCO sample, that was made and displayed the 90 K Meissner effect on January 12, showed that La123 was synthesized more than a month and half earlier. The paper was completed and submitted by express mail to *Physical Review Letters* on March 15 right before I departed for the Annual March Meeting of the American Physical Society (APS) in New York City. The results were quickly reproduced by other groups all over the world.

The "Woodstock of Physics" held in the evening of March 16, 1987 during the March Annual Meeting of the American Physics Society was a well-known extraordinary event in physics. However, I believe that few people really know its origin. It all started when I received in the second week of December 1986 a "Dear Colleague" letter (Fig. 3.12) dated December 8 from the American Physical Society (APS), rejecting our abstract for a 10 minute contributed talk entitled, "Study of Oxygen-Deficient Perovskite-Like $Ba_xLa_{5-x}Cu_5O_{5(3-y)}$ Compounds". Our abstract (Fig. 3.13), which was submitted in late November, was rejected because it "was longer than the 4 1/8 inches allowed". It was an intensely busy and exciting period of time for my group. After drafting the abstract on duplicating Bednorz and Müller's observation, I left it with one of my students and asked him to review it and make sure that it conformed to all APS rules and regulations before submission. Apparently, after reviewing it, he knew it was only about half a line too long and thought the abstract could sail through. The letter from APS put a damper on our spirits, but not for long. I felt that an exciting physics story was on the verge of unfolding. I told my dispirited colleagues, "Since we can't give a contributed talk on the early data of LBCO (abbreviation for $Ba_xLa_{5-x}Cu_5O_{5(3-y)}$) next March, let's give an invited talk instead on more exciting recent results." I immediately called Neil Ashcroft at Cornell University, who was the chairman of the Condensed Matter Physics Session of APS and was responsible for all program matters of the 1987 March Meeting. He had a long history of interest in HTS. It did not take any effort to convince him that a special session on HTS would be appropriate, after briefly informing him about the recent developments in oxide superconductors, including our high pressure work in raising the T_c. Since the entire program had been scheduled at that time, he told me that the only question was when to have it. In addition, he had to poll

"Origin of the
Woodstock of
physics" held in
NYC
3/16/87

The American Physical Society

W. W. HAVENS, Jr., EXECUTIVE SECRETARY
M. A. FORMAN, DEPUTY EXECUTIVE SECRETARY

335 EAST 45th STREET
NEW YORK, N.Y. 10017
(212) 682-7341

8 December 1986

Dear Colleague:

Your abstract has been rejected for the meeting of the American Physical Society for which you submitted it because it did not conform to the rules and regulations for submission of abstracts specified by the Executive Secretary and approved by the Council.

The rule to which your abstract did not conform is checked below:

( ) Your abstract was not submitted by a regular member of the American Physical Society in good standing.

( ) The member submitting the abstract was not one of the authors and the abstract did not include the member's name in a footnote to the abstract in the space allowed for the abstract.

( ) Your abstract was wider than the 4 3/4" allowed.

(✓) Your abstract was longer than the 4 1/8" allowed. *by the line.*

( ) Your abstract was not signed in the lower right hand corner.

( ) Your abstract did not include the name of the author or his affiliation in the abstract.

Fig. 3.12 Letter from American Physical Society rejecting abstract of Chu's paper.

Origin of The
"Woodstock of Physics"
in NYC on 3/16/87

Submitted
Nov. 1986
→ Deadline Dec 5 86

Abstract Submitted
for the March 1987 Meeting of the
American Physical Society
March 16-20, 1987

Sorting Category
22a

Study of Oxygen-Deficient Perovskite-Like
$Ba_xLa_{5-x}Cu_5O_5(3-y)$ Compounds[*]. C.W. CHU, K. FOSTER, L.

GAO, P.H. HOR, Z.J. HUANG, R.L. MENG, S.C. MOSS, L.
ROBERTSON and Z.X. Zhao. U. of HOUSTON -
Recently, possible percolative superconductivity up to
~35K was proposed by Bednorg and Müller in oxygen-
deficient Ba-La-Cu-O compounds following the detection
of a large resistance R-drop on cooling with an onset
suppressable by current. Coprecipitation from aqueous
solution and low temperature treatments were suggested
to be crucial for the observation of the R-drop.
However, by employing a non-coprecipitation technique,
we have obtained $Ba_xLa_{5-x}Cu_5O_5(3-y)$ compounds predomi-
nantly with a tetragonal perovskite structure. Some
samples with x=1 exhibit a R behavior similar to that
previously reported with a ~36 fold R-drop below ~30K.
An ac diamagnetic signal of <1% occurs at 4K. Room
temperature powder x-ray data for samples with and
without the R-drop are very similar except for two
extremely weak lines. The R-drop disappears after some
samples were exposed to air for six days, resulting in
a 10 fold increase in R. At present, the exact nature
of the R-drop in $Ba_xLa_{5-x}Cu_5O_5(3-y)$ remains unknown.
More detailed and systematic studies in sample preparation and characterization are in progress.

Fig. 3.13 Chu's rejected abstract.

members in his council before a final decision could be made. About
four hours later, as promised, Neil called back and asked me to organize
the session for one of the evening time slots that he thought he could
find. I made the initial arrangements for the special session, including
inviting the panel speakers. After their acceptance of the invitation, I
organized the panel speakers in the sequential order of their published
work on oxide HTS, i.e., Alex Müller of IBM Zürich, Shoji Tanaka of
the University of Tokyo, Paul Chu of the University of Houston,

Zhongxian Zhao of the Physics Institute of Beijing and Bertrum Batlogg of AT&T Bell Laboratories. Dramatic rise in T_c that took place within less than two months prior to the meeting turned the special session on HTS on March 16, 1987 into the "Woodstock of Physics", which I could not have dreamed of when I first proposed it. It was a moniker coined by the late Mike Schlutter of AT&T Bell Laboratories after the famous Woodstock Music and Art Fair in the summer of 1968 that drew about a half million young people in a pasture near upstate New York. The big crowd closed the New York State Thruway and created one of the nation's worst traffic jams. The "Woodstock of Physics" in 1987 clearly was smaller, but had attracted an unprecedentedly large crowd of physicists to Hilton Hotel (New York City), the conference site. Thousands of physicists packed themselves into the conference rooms and the corridors wired with televisions. Many of them stayed there discussing spiritedly HTS almost till dawn, after listening to the five panelists and more than 50 presentations by others. I indeed feel extremely blessed being able to be part of this unusual physics festival (Fig. 3.14).

Fig. 3.14 Some of the protagonists of the special session on superconductivity (March 1987, APS Meeting, Hilton Hotel, New York City). From left to right: Alex Müller, Paul Chu, and Shoji Tanaka. (Courtesy of the Niels Bohr Library, American Institute of Physics.)

3.2.4.4 *After 1987 – continued excitements*

The discovery of superconductivity in YBCO above the temperature of liquid nitrogen has ushered in the new era of high-temperature superconductivity. It has made many superconductivity applications conceived decades ago more practical, since one can use the plentiful, inexpensive, and easy-to-handle liquid nitrogen for coolong. It has opened up new frontiers for scientists to explore.

The discovery of high temperature superconductivity (HTS) in the non-intermetallic compounds $La_{2-x}Ba_xCuO_4$ at 35 K (1986) and $Yba_2Cu_3O_7$ at 93 K (1987) has been ranked as one of the most exciting advancements in modern physics, with profound implications for science and technologies. Before 1986, superconductivity was considered well understood and nothing more than a laboratory curiosity. Although several practical uses had been developed over the years, the full technological impact of superconductivity was not yet realized because of the low T_c. With the advent of HTS, the situation drastically changed. The superconductivity community was revitalized by the challenge to account for the occurrence of superconductivity at such a high temperature and the anomalous normal state properties in such an unusual class of materials. The newly discovered materials raised hope that a wide range of applications would become a reality. In the ensuing 18 years, extensive worldwide research efforts have resulted in great progress in all areas of HTS science and technology. For instance, more than 150 compounds have been discovered with a T_c above 23 K (-250°C); the T_c has been advanced to a record high of 134 K (-139°C) in $HgBa_2Ca_2Cu_3O_{9-\delta}$ at ambient and 164 K (-109°C) under pressure; many anomalous properties have been observed; various models have been proposed to account for the observations; and numerous prototype devices have been made and successfully demonstrated. In spite of the impressive progress, the mechanism responsible for HTS has yet to be identified; a comprehensive theory remains elusive; the highest possible T_c is still to be found, if exists; and commercialization of HTS devices is not yet realized. My group continues to work hard on the discovery of new HTS and related materials, understanding the occurrence of HTS and the use of HTS for practical purposes.

3.2.5 *University presidency*

Life is a strange encounter. In 2000, a difficult-to-refuse opportunity dogged me repeatedly and finally has helped expand my horizon from that of a scientist in the trenches at the University of Houston (UH) to that of the President of Hong Kong University of Science and Technology (HKUST) (Fig. 3.15) in the Orient, a young and rising premier university in the most dynamic region of our 21st century world. The vision, generosity, and hard work of colleagues on both sides of the Pacific since 2000 have demonstrated once again that exceptions do exist to the expression "one cannot have his cake and eat it too." Science in our Houston lab has continued to flourish in the past four years, while new heights have been scaled by HKUST on her way to becoming a world-class university. Synergistic collaboration is being developed between UH and HKUST in the area of the almighty nanoscience and technology.

Fig. 3.15 Dr. Sze-Yuen Chung, the Pro-Chancellor, installed President Chu at Hong Kong University of Science and Technology.

Chapter 4

Eli Ruckenstein: Leader in Chemical Process Development

4.1 Introduction by the Editor

4.1.1 What is chemical process development?

A chemical process refers to a process that involves the chemical change of one or more substances. These processes pertain not only to chemical reactions, but also to the transport of the chemical substances involved, the formation of a mixture of different substances (e.g., formation of a suspension of fine particles in a liquid medium), the separation of a substance from a mixture of various substances, the control of the rate of a process (e.g., control through the use of a catalyst), the degradation of substances during use, the modification of substances (e.g., by coating) for improved performance, the management of chemical wastes, the protection of the environment, and the manufacture of products in gas, liquid and solid forms. Chemical processes are central to the chemical industry, whether the products are detergents, paints, drugs, skin cream, toothpaste, polymers, batteries, filters, etc. The development of a chemical process that is of practical importance requires knowledge of both chemistry and engineering (a branch of engineering often referred to as chemical engineering), although physics, biology and mathematics can also be involved.

(a) Age 19   (b) Age 42   (c) Current

Fig. 4.1 Dr. Eli Ruckenstein. Over the years, then and now

4.1.2 Scientific contributions of Dr. Ruckenstein

Dr. Ruckenstein (Fig. 4.1) is an international leader in chemical process development. His extensive research in a period of over 50 years has resulted in enhanced understanding of the mechanisms of chemical processes, and the development of improved processes and materials. His work is of profound scientific significance, as his findings have pushed the frontiers of scientific understanding in relation to many types of chemical processes. In addition, his work has far-reaching effects on the chemical, medical, environmental and electronic industries. Specifically, Dr. Ruckenstein has contributed to the understanding of the transport and separation of substances, the development of catalysts, colloids and emulsions, the alleviation of biological problems (such as the problem related to the reaction of blood with the surface of an implant), and the process of making polymers (i.e., the process of polymerization).

4.1.3 Honors received by Dr. Ruckenstein

Dr. Ruckenstein has received a large number of awards for his outstanding research (Fig. 4.2). These awards include Founders Award from U.S. National Academy of Engineering (2004), Founders Award from American Institute of Chemical Engineers (2002), U.S. National Medal of Science (from President Bill Clinton in 1998, Fig. 4.3), Pioneer

Fig. 4.2 Dr. Ruckenstein Being honored in recognition of exemplary contributions to research and scholarship by the State University of New York in Albany, NY, on October 20, 2003. Left to right: Dr. George Stefano, Vice Chair of the Board of Directors of the Research Foundation of the State University of New York; Dr. Ruckenstein; Dr. Jaylan Turkkan, Vice President for Research, University at Buffalo, State University of New York; Mr. Randy Daniels, Vice Chair of the Board of Trustees of the State University of New York.

Fig. 4.3 Dr. Ruckenstein receiving the U.S. National Medal of Science presented by President Clinton in 1998.

of Science Award from Western New York, Hauptman-Woodward Medical Research Institute (2002), Chancellor Charles P. Norton Medal from State University of New York at Buffalo (1999), E.V. Murphree Award in Industrial & Engineering Chemistry from American Chemical Society (1996), Langmuir Distinguished Lecturer Award from American Chemical Society (1994), The Jacob F. Schoellkopf Medal from the Western New York Section of the American Chemical Society (1991), The Walker Award from American Institute of Chemical Engineers for Excellence in Contribution to Chemical Engineering Literature (1988), The Kendall Award from the American Chemical Society for Research in Colloids and Surfaces (1986), Creativity Award from U.S. National Science Foundation (1985), Senior Humboldt Award from Alexander von Humboldt Foundation (1985), The Alpha Chi Sigma Award from the American Institute of Chemical Engineers (1977), George Spacu Award from Romanian Academy of Science (1963), and The Romanian Department of Education Award for Teaching (1961). Dr. Ruckenstein was elected a member of the U.S. National Academy of Engineering in 1990.

4.1.4 Career development of Dr. Ruckenstein

Dr. Ruckenstein was born in Romania and received his Ph.D. degree from Polytechnic Institute in Bucharest, Romania. During 1949-69, he was Professor at Polytechnic Institute. During 1970-73, he was Professor at the University of Delaware, USA. Since 1973, he has been on the faculty of University at Buffalo, State University of New York, where he is currently Distinguished Professor in the Department of Chemical Engineering.

4.2 Dr. Ruckenstein's Description of His Life Experience

I was born in Botosani (Fig. 4.4), a small town in northern Romania, located in the hilly area of Moldavie (or Moldova) (Fig. 4.5) between the Carpathian Mountains (Fig. 4.6) and the river Pruth, or Prut River (Fig. 4.7), which originates in the Carpathian Mountains in Ukraine, flows

Fig. 4.4 The town of Botosani.

Fig. 4.5 Map showing Moldavie and its neighboring regions.

Fig. 4.6 The Carpathian Mountains.

Fig. 4.7 The Pruth River.

southeast to join the Danube River, and forms the border between Romania and Moldova. Botosani is not too far from the border with the former Soviet Union (a short name for the Union of Soviet Socialist Republics, or USSR, which was a state in much of the northern region of Eurasia in 1922-1991). The city was an agricultural center populated with many *boieri* (landowners). The numerous *paysans* who farmed the land for them lived in humble shacks, huddled together in small villages. Botosani itself was divided into two sections: one containing the magnificent homes of the *boieri* while the other was a slum, populated largely by Jews. While the novels of Gogol and Turgheniev provide an image of the opulence of the *boieri*, the writings of Shalom Aleihem describe the misery of the *luft menchen* living in those slums.

My father was relatively prosperous when I was born, but our financial situation deteriorated and became desperate during the great depression of 1930. Few memories from my childhood deserve to be mentioned. I refused to learn the alphabet until my teacher read for me a few chapters from *Cuore* (Heart) by Edmondo d'Amicis (Italian writer, 1846-1908) and a few chapters from *Robinson Crusoe* (1719 English novel by Daniel Defoe). From that moment on my fascination with books has only increased.

Our education at the lyceum was based largely on the French tradiation; yet I was attracted to the Russian novelists. When I was 12, I discovered Gogol (Ukrainian-born Russian writer, 1809-1852) and Dostoievsky (Russian writer, 1821-1881) and not much later, Tolstoi (Russian writer, 1828-1910) and Gorki (Russian writer, 1868-1936). My adolescence was strongly influenced by these writers and I could hardly wait to return home each day to continue reading. I am not a loner but my fascination with Lermontov (Russian writer, 1814-1841), Pushkin (Russian writer, 1799-1837) and Dostoievsky (Russian writer, 1821-1881) transformed a child into an adolescent, perhaps too early.

I never visited Russia. The only message from that country was the cold wind, Crivatul, which came from the Russian Steppes (temperate grassland of Eurasia, consisting of level, generally treeless plains) each winter. But to me, this wind was bringing the warm humanitarianism of the Russian novelists. I am disappointed by the modern Russian literature. The policy of the Communist Party has destroyed a great cultural tradition.

When I was 14, the Second World War started and racial laws eliminated me from the state schools. The Jewish community organized a private school in which very few of our instructors were professional teachers. They were intellectuals who had lost their jobs because of the racial laws, but they had unusual enthusiasm, dedication, and that strength of character built by difficulties. One of them, Miron Grunberg (1921-1943, sent in 1941 to force labor), was only a few years older, but he had an unusually broad culture and intelligence. He was a model for all of us. Because of forced labor, he died while still young. I was glad to learn, two years ago, that his volume of poetry has been published in Israel by his mother. Having achieved this goal, she died a few days after its publication.

At seventeen I was drawn into forced labor and worked on the construction of buildings by carrying bricks onto a scaffold. Despite the long hours, I stubbornly prepared for exams for the 11th and 12th grades and although I was out of school, managed not to fall behind in my studies. The Red Army liberated Botosani from the Fascist occupation (Fig. 4.8) in April, 1944. Later that year I moved to Bucharest (the capital of Romania), hoping to be accepted at the University.

I had no idea what to expect from a University or even what career to pursue. Having grown up in a small town, I felt completely lost when I

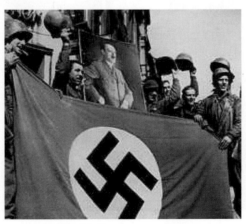

Fig. 4.8 Fascist occupation.

first arrived in the metropolis of Bucharest. Following the advice of a family friend, I finally decided to compete for a place in chemical engineering at the Polytechnic Institute. I was the first person in my family to go to college.

Life as a student was a constant struggle. Inflation soared: my father's monthly salary was but enough to buy a bus ticket from the University to the place where I was living. Hunger, cold and humiliation were not strangers to me.

There were also happy occasions. In 1946 I met Velina Rothstein and we became close friends. Together we read Spinoza and Nietzche. Even now I marvel at the unusual culture of this girl, then 16. We were married in 1948. No one has influenced my life more than Velina (Fig. 4.9).

The Chairman of the Department of Chemical Engineering at the Polytechnic Institute was Emilian Bratu (founder of the Romanian School of Chemical Engineering), one of the most admirable men I ever met. After five years of study and my graduation in 1949, he chose me to be his assistant and provided me with the best possible conditions, under the circumstances, for research. As an assistant professor, I aided Bratu in recitation and laboratory sections of his courses.

Fig. 4.9 Dr. Ruckenstein and his wife, Velina.

By tradition, there were no formal graduate courses offered by the Institute. The library had very few of the books that were available to graduate students in the West and received journals only after a year or more of delay. So *Industrial and Engineering Chemistry* became my graduate school, my textbook, and my teacher. I read each volume in series, some from cover to cover. Although there are some advantages to this procedure, it is most inefficient and is not to be recommended. It is like trying to deduce the plot of a novel by starting in the middle. Without knowledge of earlier developments, I began to understand the papers only through stubbornness and repetition.

In addition to a qualifying examination and a dissertation, a major requirement for the Ph.D. degree was to pass a test on Marxism-Leninism. Some of my colleagues spent more than a year studying for this exam. I chose not to take it. In 1966 the law was changed to eliminate this requirement and then, at the urging of my wife, I submitted some of my papers as a dissertation and received the Ph.D. degree.

The early days of my career coincided with the early days of communism in Romania. I was not a Party member, and I did not adjust easily to the very political tone of the times. In addition, I worked in a scientific community that had few, if any, connections to the outside world.

In Romania, as in most communist countries, financial support for research was given to a small group, chosen more or less arbitrarily by the Party. Usually, assistant and associate professors worked for the department chairman, who then became well known because his name was routinely added to any publication submitted from the department. Some of these chairmen were distinguished scientists and deserved recognition; many others acquired their position more through Party membership than scientific competence.

Professor Bratu, who is no longer alive, was unusually fair and modest in this respect. He provided academic freedom and refused to sign his name as an author on a paper when he felt he had not contributed enough. With the passage of time my respect for him continues to grow as I better appreciate how much I benefited from the unusual atmosphere he provided.

Understandably, papers published in Romanian, or even Russian, journals are largely unknown here. Until late 1950, we were not allowed to send manuscripts to the West. When the ban was finally lifted, manuscripts were returned due to "lack of space." During the McCarthy era, scientists paid twice for living under a communist regime.

In the atmosphere of scientific exchange which followed President Nixon's goodwill trip to Romania in August, 1969, I was allowed to come to this country as a senior NSF (National Science Foundation, USA) scientist at Clarkson College of Technology. Velina accompanied me but our two teenage children, Andrei and Lelia, stayed in Romania with Velina's parents. I spent a most interesting year at Clarkson, interacting with the faculty.

From 1970 to 1973 I was at the University of Delaware. In retrospect, I perceive this to have been a period of change in my research direction. Published accounts of scientific research are much more readily available in the U.S. than in Romania. I became overwhelmed by the amount of information available and the desire to digest all of it. With this new knowledge came new questions, which I then tried to answer and in the congenial atmosphere at Delaware I had many enriching discussions with my colleagues. Gradually my research shifted from heat and mass transfer to catalysis and colloids, but only with the benefit of time does this change appear as a discontinuity.

At that time I was too preoccupied with the frustration of trying to get our children out of Romania to notice the change. It was a very difficult period for Velina and me. We shall never forget the moral and material support we received from Arthur B. Metzner (H. Fletcher Brown Professor of Chemical Engineering) and others at Delaware. Finally, after two years of persistent effort and a lot of luck, my family was once again together, this time in the U.S.

Over the course of my career, I worked on a broad range of problems, many of which are conventionally associated with subfields of physics, chemistry and engineering. Engineering research is a synthesis, which implies learning whatever experimental or theoretical techniques are required to solve the problem at hand, many times forays into distant disciplines.

My own way of doing research was to a large extent defined by the fact that I am basically self-taught and that I do not belong to any school. I was educated during a special time in the history of Romania, my native country, immediately after the Second World War. A tradition in pure mathematics had existed for some time, and physics and chemistry benefited from the return of a few distinguished researchers who had been trained in the West. However, Romania had no tradition in modern engineering, and much of my education consisted of studying the available original literature on my own.

My early work on heat and mass transfer - which did not get to the West until 1958 - owes much to the encouragement and understanding of my teacher and mentor, Professor Bratu. In spite of uncertainties under the Communist regime and the fact that I had no formal authority, I managed to attract a group of somewhat younger colleagues with whom I cooperated on a variety of problems, from mass and heat transfer to kinetics of gene expression to interfacial phenomena and thermodynamics of small systems.

When I arrived in the United States at the age of 45, everything I had learned about the West was from a limited number of journals and an even more limited number of contacts with researchers from outside Romania. I was immediately struck, overwhelmed by the amount and breadth of information that became available to me overnight. At that point, I really started a new career.

Since then, my work has extended in roughly five different directions: catalysis, colloids, separation of proteins, polymers, and material science.

In the area of catalysis, my students and I studied theoretically and experimentally the stability of small metallic clusters on catalytic supports. The other area of catalysis I was fascinated with was the kinetics of selectivity of supported catalysts. I also formulated a theory for the mechanism of oxidation by mixed oxides.

In the area of colloids, we developed a hydrodynamics of colloidal particles that accounted for double-layer and van der Waals interactions. We introduced the concept of "interaction force boundary layer." We also proposed thermodynamic theories of surfactant aggregation, microemulsions, and liquid crystals. I feel happiest about two contributions: (1) the reformulation of the classic theory of double-layer forces to include in a unified way the interplay between double-layer, hydration forces and the hydration of ions; and (2) a unified kinetic approach to nucleation and growth.

Rather than boring you with descriptions of other theoretical efforts, I will mention some of our more recent technological innovations: the development of a solid-solution catalyst for carbon dioxide (CO_2) reforming of methane; the preparation of some interesting compounds for the storage of hydrogen (H_2); and the preparation of a paste with high thermal conductivity, which is now used in all IBM computers. In addition, from concentrated emulsions we developed various technologies for preparing polymers, conductive polymers, and membranes for separation processes.

My career has been and remains a great source of satisfaction to me. For this, I must acknowledge my students, postdocs, and collaborators who, through their hard work, patience and dedication, have taught me, inspired me, and stimulated me.

I consider myself a very lucky man who has been surrounded by many guardian angels. I would not have survived my early days as a young assistant professor at the Polytechnic University in Bucharest without Professor Bratu, who protected me from my own inability to fit into the politically driven academic environment of communist Romania, where I was considered a dangerous reactionary.

I also want to thank E. James Davis (Fig. 4.10(a)), William N. Gill (Fig. 4.10(b)), and others at the Chemical Engineering Department of Clarkson University, who were responsible for my initial move to the United States. I owe them all a debt of gratitude.

I also owe a debt of gratitude to Professor Arthur B. Metzner (Fig. 4.11) and other colleagues at the University of Delaware, who welcomed me and my family warmly, made us feel at home, and helped me through the years of adjustment to the American way of life and the American academic system. I will never forget their moral support for me and my wife during the two-and-one-half years we struggled to get our then teenage children out of Romania.

I also want to acknowledge Professors George C. Lee (Fig. 4.12(a)), William N. Gill (Fig. 4.10(b)), Ralph T. Yang (Fig. 4.12(b)), Carl R.F.

(a) Professor E. James Davis   (b) Professor William N. Gill
Fig. 4.10 My colleagues at Clarkson University.

Fig. 4.11 Professor Arthur B. Metzner of University of Delaware.

(a) Professor George C. Lee   (b) Professor Ralph T. Yang   (c) Professor Carl R.F. Lund
Fig. 4.12 My colleagues at University at Buffalo, State University of New York.

Fig. 4.13 Professor Howard Brenner of Massachusetts Institute of Technology.

Lund (Fig. 4.12(c)), and other colleagues at University at Buffalo, State University of New York for providing the resources and helping create the supportive atmosphere that made my last 25 years the most productive, enjoyable and rewarding of my career.

Someone who deserves a special mention is Professor Howard Brenner (Fig. 4.13), who was instrumental in my getting my first position in the United States. I am deeply thankful to Howard for his help, encouragement, and friendship over the years.

Finally, I would like to thank my family, especially the person who has been the greatest influence in my life, my wife Velina, for her constant support, love, and selfless dedication during our 57 years together. Without her, I would not have the wonderful children I have, I would not have as many friends as I have, I would not have had the career I have had, and I certainly would not be here today.

Science has been my life, and I have been lucky to have the opportunity to wake up every morning excited about my work and curious about the next adventure. On my next birthday, I will turn 80, and I still hope to find that intangible holy grail.

I started my career in those dark days in Romania with little but ambition and youthful exuberance on my side. I now live in a free country, and I still have the ambition - and on a good day a little exuberance.

Chapter 5

Jennie S. Hwang: Pioneer in Surface Mount Technology and Environment-Friendly Lead-Free Electronics

5.1 Introduction by the Editor

5.1.1 What is surface mount technology?

Surface mount technology (SMT) is considered to be the backbone manufacturing technology of the miniaturization of electronics, making today's gadgets possible, be it medical device, notebook computer, cell phone, digital camera or DVD music player.

The impact of surface mounting on printed circuit assembly is dramatic and profound. It is one of the most significant developments in the electronics era during the last two and a half decades. SMT embraces the advancement of semiconductors and the electronic packaging, leading to the end-results of fast speed, decreased size/weight, and increased functionality/overall improved performance.

5.1.2 What is environment-friendly lead-free electronics?

Environment-friendly electronics has evolved to a global commitment.

As the miniaturization and functional convergence continues, IT (information technology) gadgets and/or electronic/optical products will continue to drive technological innovations. Solder, albeit a small percentage in this immense industry, is a niche area and plays a crucial role, due to its critical functionality, as the electrical, mechanical and thermal linkage of electronic circuitry.

Solder joints are numerous in an electronic circuit board that runs a piece of electronics, so their performance and reliability are mission-critical. The lead (one of the most toxic substances) in a conventional solder alloy causes environmental concern. Thus, the development of environment-friendly lead-free solder material for achieving environment-friendly electronics is one of the hottest areas of research in the field of electronic packaging.

Today, the delivery of environment-friendly lead-free electronics is meeting the global commitment.

5.1.3 Honors and recognition received by Dr. Hwang

Dr. Hwang's wide-ranging career encompasses international business, corporate executive, CEO of three start-up companies, invention, worldwide manufacturing and technology services, leadership positions of non-profit organizations, as well as corporate and university governance. Among her many awards and honors are citations by the U.S. Congress and the Ohio Senate/House for outstanding achievements; election to the U.S. National Academy of Engineering; induction into the Women in Technology International Hall of Fame (Fig. 5.1); induction into the Ohio Women's Hall of Fame; the recipient of Founder's Award of the Surface Mount Technology Association; and being named a "Star to Watch" by *Industry Week*. She has held various "woman pioneering" capacities. At her YWCA Women of Achievement Award, her citation read: "Being honored as the FIRST WOMAN is a way of life for Dr. Hwang..., including being the first and only woman holding the national presidency of the Surface Mount Technology Association (a U.S.-based professional organization comprising university, government and industry members), the first woman to receive the Ph.D. in Materials Science & Engineering from Case Western Reserve University ..."

She also received Distinguished Alumni Awards from Case Western Reserve University and National Cheng-Kung University (Figs. 5.2 and 5.3), and the Special Achievement Award from Kent State University. In addition, she was elected to Ohio Commodore (Fig. 5.4), and featured in *"Women in Chemistry"* of Chemical Heritage Foundation; the "Women in Technology" of American Society of Mechanical Engineers; the "Gallery

Fig. 5.1   Fig. 5.2

Fig. 5.3   Fig. 5.4

Fig. 5.1 Induction into International Hall of Fame—Women In Technology (with the other 3 inductees—Drs. Shirley Jackson, Bonnie Dunbar, Irene Greif)

Fig. 5.2 Receiving the Distinguished Alumna Award at ChengKung University

Fig. 5.3 Acceptance speech of Distinguished Alumna Award at ChengKung University

Fig. 5.4 Induction into Ohio Commodore (with Governor Bob Taft)

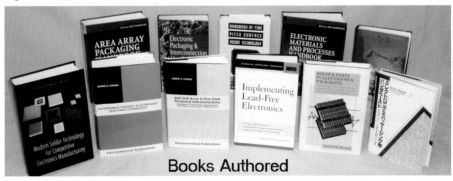

Fig 5.5 Books authored over the years

of Women Engineers" of National Academy of Engineering; "Role Models for Women in Engineering, Technology" of WEPAN (Women in Engineering Programs & Advocates Network) and the Crain's Business magazine: "Technology Leaders".

5.1.4 Contributions of Dr. Hwang in technology and scholarship

Dr. Hwang is an internationally renowned authority in SMT technology, and a major contributor to the worldwide infrastructure establishment of SMT manufacturing. She has extensive hands-on experiences in technology-transfer and bringing innovations to commercialization.

As an inventor of a number of patents, she is the author of over 250 publications, including being the sole author of several internationally-used textbooks. She is also a co-author of several books on the subjects related to the leading technologies. As a columnist for *SMT* (a globally circulated trade magazine), she addresses technology issues and global market thrusts monthly; her views are widely solicited and highly regarded worldwide (Fig. 5.5).

In addition, she is a prolific author and speaker on education, workforce, and social and business issues.

As an invited keynote speaker, she shares her thoughts and vision with various institutions. Her topics range from commencement keynote speeches (Fig. 5.6) at universities to the emerging technologies at the U.S. Patent and Trademark Office (USPTO). Over the years, she has taught over 15,000 professionals and researchers in professional advancement courses, focusing on disseminating new technologies and providing professional advancement education to the workforce. She is a popular keynoter and featured speaker at the national and international events, including the annual meeting of American Association of University Women—Cleveland, American Society of Mechanical Engineers—Electronic Division, Scandinavian Electronics Packaging & Production Conference (Sweden), Japan Opto-Electronics Packaging and Production meeting (Tokyo, Japan), The Electronic Packaging & Production Conference in Shanghai (China), The American Material Society, Intersociety Electronic & Photonic Packaging Conference (Hawaii), Annual meeting of Electronic Circuit Board Assembly

Fig. 5.6 Fig. 5.7

Fig. 5.8 Fig. 5.9

Fig. 5.6 Commencement speech at Kent State University
Fig. 5.7 Featured speaker at an industry event
Fig. 5.8 Invited lecture on emerging technology in Hong Kong
Fig. 5.9 Featured speaker at the International Conference on Women in Technology

Fig. 5.10 Appointed to the U.S. Commerce Department's Export Council

(Australia), Annual Conference of International Microelectronics and Packaging Society—Brazil (Campinas, Brazil), Annual Conference on Electronics Circuit Assembly--Singapore (Singapore), and numerous industry events. She is an invited lecturer across the US and abroad, including Israel, France, Germany, Belgium, Sweden, England, Hong Kong, Singapore, Malaysia, Taiwan, China, Japan, Puerto Rico, Brazil, Australia, etc…(Figs. 5.7-5.9)

5.1.5 Contributions of Dr. Hwang in business

In 1990-1994, Dr. Hwang was Co-Founder and President/CEO of International Electronic Materials Corp., a material manufacturing company that serves the worldwide electronics industry. The company was acquired by a suitor. Afterwards, she founded H-Technologies Group, Inc., an intellectual property/service company that focuses on innovation, licensing, partnership and technology transfer. She concurrently serves as the Interim CEO for Asahi America, Inc.

Prior to founding her companies, Dr. Hwang's corporate life covered several senior executive and research positions with Lockheed Martin (Martin Marietta Corp.), Hanson PLC. (SCM Corp.), and Sherwin Williams Co.

Dr. Hwang has served in the governance capacity for Fortune 500 NYSE-traded corporations, non-profits and universities, including on the board of Ferro Corporation, Second National Bank, Asahi Chemical, Case Western Reserve University, Cleveland State University Foundation, ASM International, SMT Journal (Britain), National Electronic Production Conference/Exhibitions, Kent State University – College of Arts and Sciences, and Singapore Advanced Manufacturing Technology Institute. In 1994, Dr. Hwang was elected as National President of Surface Mount Technology Association, an organization that provides education, training and networking opportunities for university, government and industry members.

Furthermore, she has managed complex consulting projects and consulted to many Fortune 200 companies and to U.S. Government Programs.

5.1.6 Contributions of Dr. Hwang in civic and professional services

The influence of Dr. Hwang goes beyond technology and business. She has served as a trustee of Cleveland Council on World Affairs and of San Jose/Cleveland Ballet. In addition, she was President of Organization of Chinese-Americans (Greater Cleveland area) and was a board member of Great Lakes Science Center and the Cleveland Museum of Art—Technology Committee. She is also appointed to the U.S. Commerce Department's Export Council (Fig. 5.10) and served as the co-chair of Government Relations and Public Policy Committee of IPC.

She is a reviewer of various publications and books and served as judge for the Annual R&D 100 Award, Innovative Product Award, Women Business Owner Association's Annual Award, and the Heinz Awards. She often presides, chairs and moderates various industry and professional committees and conferences.

In addition, she served on the National Research Council/National Academy of Engineering – Diversity Forum and the R&D Globalization Committee.

For the last ten years, she has been heavily engaged in university activities, ranging from serving as a Trustee of Case Western Reserve University to chairing the Arts and Sciences College Advisory Board of Kent State University to serving on Case Western Reserve University's Presidential Search Committee.

In support of higher education, science and engineering, Dr. Hwang established a YWCA Annual Award recognizing outstanding women students who study in science, engineering. She also established an endowment at Case Western Reserve University, designated to encourage faculty and students to acquire international exposure. Additionally, a Faculty Excellence Award was set at Cleveland State University, honoring faculty's exceptional research and service performance.

5.1.7 *Formal education of Dr. Hwang*

Dr. Hwang received a Ph.D. degree in Materials Science and Engineering from Case Western Reserve University (USA), an M.S. degree in Chemistry from Columbia University (USA), an M.S. degree in Liquid Crystal Science from Kent State University (USA), and a B.S. degree in Chemistry from National Cheng-Kung University (Taiwan). In addition, she attended various management and leadership programs and the Harvard Business School's Executive Program.

5.2 Dr. Hwang's Description of Her Life Experience

Nothing would have happened without the nurturing, encouragement and support of my family—to learn, study, excel and then reach for the stars.

It has been an intensive and versatile experience. One eternal challenge is time management and priority setting.

To me, life has been a progression of three stages: the first, schooling and preparing for the future by rooting a fundamental education into a solid base of learning and experience from which to grow. The second stage was intellectual growth and seasoning, continued learning, experiencing and contributing. The best for last is the third stage, which I consider the prime of my life, it is the time when I can be most productive and valuable. It is the time that I view that I can contribute the most with what have been learned.

Working in full intellectual vigor is always my joy.

5.2.1 *The first stage – schooling and preparing for the future*

Growing up in an intellectually stimulating family, my grandfather had high aspirations for me. Schooling was of ultimate importance. However, in addition to strictly following the formal curricula, I was encouraged to participate in extra-curricular activities, such as sports, calligraphy, singing, dancing, etc. (Fig. 5.11). I have special memories of playing chess with my grandfather and of his teachings in calligraphy — an activity for which I earned awards in high school (Fig. 5.12). I had a precious, rare opportunity to be showered by all his intellect. For that,

Fig. 5.11   Fig. 5.12

Fig. 5.11 Fashion and modeling beckoned the young Jennie

Fig. 5.12 Poses with her award-winning calligraphy, an art her grandfather taught her. The lessons began when she was 3 years old and ended when she was 15

I will be eternally grateful and cherish forever. There is value in how a family can affect a young person's thinking and development. I firmly believe that my thoughts reflect positive behaviors and style of encouragement used by my parents and grandfather during my formative years, as well as the inner discipline they instilled within me.

During those formative years, my family made me feel that there was no gender inequity in terms of learning or opportunity. I never had the sense that girls should not have ambition or a career goal. Looking back, I was truly fortunate.

All my teachers contributed to my education and intellectual development. However, I do recall a teacher from third grade who wrote on a poster that I was a star student and going to be an eminent leader. I was so young and didn't really understand what that actually meant, but I was so motivated by his praises…I wanted to do everything to please all.

I gravitated to topics such as science and engineering. With chemistry education in college, I pursued graduate studies in liquid crystal, a high profile field in the 1960s. The one school with a main track curriculum in this area was Kent State University, Ohio. At Kent State, I recall that Professor Gould challenged the students through critical questions. He was demanding; he challenged us in how we thought and spoke. Today the fact that I particularly welcome the challenging questions and love critical and strategic thinking may well be attributed to his demands on his students then. Upon completion of my studies in liquid crystal science, I went to Columbia University, earning my third degree, an M.A. in chemistry, before returning to Ohio. This brings my recollection of another professor, Dr. Arthur Heuer, who strongly urged my pursuit of Materials Science and Engineering. The transition from science to engineering made me nervous initially, since some terminological terms were so foreign to me. So I studied very hard, and it turned out that I received the highest grade of those engineering graduate courses in the class. This experience reaffirmed to me that effort does work.

At Case Western Reserve University (CWRU), I received a Ph.D. degree from the engineering school, where, at that time, female students were a rarity. I was the first woman to earn a Ph.D. degree from CWRU in Materials Engineering. (Two decades later, I serve on the Board of Trustees of CWRU, offering a different set of perspectives and contributions.)

With both science and engineering training, I found it tremendously facilitating in my subsequent endeavors in the business world as well as in the technology enterprise.

By the time I was about to complete my formal education (non-stop from kindergarten to Ph.D.), I had acquired a fascination for business. I found the decision and strategy-making process challenging and intriguing; meanwhile, I enjoy the tangible results of science and engineering. I had a tough time picking a career between business and academia. I could not make a distinct separation in my interest to go either track, so I decided to embark on both. I have tried to work three times harder in order to cover both.

In recent years, I have shared my experiences with young people. After one gathering, I received a note, an excerpt follows: *"Dear Dr. Hwang: We would like to sincerely thank you for the time you spent with us last Thursday. Your willingness to share information about your career and what skills and abilities were necessary to build your career was greatly appreciated... We particularly enjoyed discussing why you chose to integrate both academia and industry, and how you developed your career. We also enjoyed learning about your educational background, and how it has prepared you in the role you play today... Your perspective and advice given for women pursuing a leadership position were very helpful....".*

5.2.2 The second stage – intellectual maturation and growth

This was the time for a major shift, striding from the non-stop schooling to the workplace.

My first career choice was to enter industry, and that first job at Martin Marietta (now Lockheed Martin) was truly a learning experience. I was hired as a scientist. With hard work, I moved up the corporate ladder within two years to a senior management position. I was again the first woman working in that corporate environment—one of a male-dominated industry.

At this time, I indeed felt the gender distinction.

Moving up the corporate ladder, what I learned was the art of managing the intricate and subtle balancing among performance, intellect, ambition, assertiveness and femininity. Learning to be assertive, but in a non-threatening way, and maintaining a feminine persona was a challenge. (These feelings and observations continue, not only in the workplace but also when serving on committees, councils and boards.)

My work was never the traditional 9 to 5 route, it was indeed and continues to be that saying—the work goes where I go.

During my time at Lockheed Martin, I was responsible for advancing the coating products and the chemistry and curing process of cement-based materials used in heavy construction—bridges and highways. The ultimate objective was to improve performance and durability. At SCM Corp., I was responsible for entering the fast-moving and highly competitive electronics industry that was in revolutionary change. The days of a computer filling an entire room were becoming passé—and I was right there at the start. Recruited as a director for the new division, I had the opportunity to try out an "intra-preneurship" within the corporation, starting from almost scratch. I was responsible for the gamut of functions of research, product development, budgeting, marketing, global positioning, internal and external sales—national and international in scope—with travel and exposure to numerous industrial facilities throughout the world. Chairing both marketing/sales and technology/product task forces ignited a further sense of full responsibility. When I was called the <star performer> by our corporoate executive, all my cylinders fired up. As a material supplier to enter into a fast-moving new market, there was no shortage of challenges. At my boss' retirement, he wrote: *"Dear Jennie: ...I enjoyed much our vigorous discussions, from which I learned much. Your unique talents made our product line possible, which was an extra pleasure for me,...."*—a great feeling in reading this note. Customers then included all household names of the era. That was a real exciting time in terms of new technologies and the demands of supporting a whole new industry. It was like I was starting up a whole new business.

During this time, in addition to contributing to several co-authored books, I authored my first textbook on the Technology and Application for Electronics Packaging in 1988. Writing that first book was personal and somewhat overwhelming—taking much of my personal time on weekends, evenings and holidays. But it was the most relishing journey, and I learned a lot during the process.

5.2.3 The third stage – a time to contribute with what has been learned

The desire for full accountability and challenge coupled with the worldwide recognition gained from the release of my first book made the entrepreneurial endeavor most appealing. In 1990, driven by an internal vision that saw the market demand and an entrepreneurial opportunity, I ventured to form the first of three businesses that I would create in the last fifteen years. International Electronic Materials Corp., formed to develop technologies that would produce better materials and enable circuitry to fulfill miniaturization demands in the global market place. This first entrepreneurial business was truly started from scratch. It was a manufacturing company with the first product line to supply interconnecting solder materials to make circuit boards of ever-smaller, faster electronics. While the challenge was keeping up with the rapid pace of technological evolution, I found it even more stimulating running the business and keeping up my hands-on persona. Within four years, the company held a position of worldwide sale network of commercial products. A successful potential with an international base led to a welcome acquisition in 1994. When approached about selling the company, I made a business decision to sell—but with my eye to the future.

Envisioning the emerging new business model in "core competency vs. outsourcing," in 1994, I started H-Technologies Group Inc., an intellectual property and service company that innovates and provides services to businesses throughout the world.

One of the high-impact projects is to help the industry globally, to implement environment-friendly electronics, to remove any use of what is considered to be most hazardous materials, such as lead, and to make a lead-free type of electronics.

Under the U.S. Defense Mantech Program and U.S. Army Materiel Command with the goal of reducing the cost and enhancing the reliability of electronic weapons on a national scale, improving the scientific basis of process controls used on production lines (focusing on the soldering technology thrust area) to advance material properties and meanwhile to address the environmental concern (to make Lead-free) was one of the missions. I was invited to participate as an advisor for the program.

This was an initiative and a start. But the Lead-free issue had been through ups and downs over the years in the global landscape. Some denied the effort at the outset, and many stopped the effort at some point. Nonetheless, our team has never wavered, maintaining a steadfast and relentless effort on the materials research and the manufacturing know-how development, harvesting patents, licensees and commercial products.

After fifteen (15) years, today, achieving environment-friendly electronics has become a reality and a global movement. I feel that our sustained effort is paying off.

I have ventured to share the information and knowledge and to put them into an integrated form. As the result, two books were published. I have responded to as many invitations as my time permits in lecturing on the subject across U.S. and abroad.

The first lead-free book, also my fifth book, *"Environment-Friendly Electronics— Lead-Free Technology,"* focusing on technology was published by Electrochemical Publications, LTD., Great Britain in 2000. And then the second lead-free book, also my sixth book, *" Implementing Lead-free Electronics—A Manufacturing Guide"*, focusing on electronics manufacturing know-how was released in 2005 by McGraw-Hill.

5.2.4 Work and family

"Work and family" is an issue close to my heart. As a mother of two children, I can speak on this with personal experience and from the bottom of my heart.

I believe that marriage requires a reciprocal support and mutual respect, and that parenthood involves sacrifices. It has always been my principle and practice to put my husband's work and children's needs before mine, regardless of how busy I was. Aside from work, my time was for my children. (Figs. 5.13-5.16)

In a way, I have managed my career around children and family, for example, choosing to be based in the Cleveland area, although extensive travel was necessary. But I had made the commitment not to move around until the kids were grown.

I still vividly remember how hard it was to leave home going on business trips when my children were young. At this point, I would like to share some of my daughter's notes, when she was little, that I have saved over the years. Every time I was on a trip, our daughter always "drew" a note and stuck it in my brief or luggage. Here is one: *"Dear Mom, I hope you have a nice four days in California. I am really going to miss you. I hope you have a very safe trip in California. I really love you and I hope you have a good presentation and have a lot of people there. Love, Lindi, P. S. Thank you for trying to get me the porcelain doll!"* ....

"Dear Mom, I hope you have nice time in Singapore. I'm going to miss you for a whole week. I love you very much. I hope you have a good meeting. And also a safe trip there and back.

Fig. 5.13   Fig. 5.14

Fig. 5.15   Fig. 5.16

Fig. 5.13 The family pauses for this family photo
Fig. 5.14 On a family vacation
Fig. 5.15 Son and daughter
Fig. 5.16 Family dog and daughter in our yard in a Sunday afternoon

Fig. 5.17   Fig. 5.18

Fig. 5.17 Son Raymond's graduation from Harvard University's Medical School
Fig. 5.18 Daughter Rosalind's graduation from Wellesley College

Try to call us every day. Will you get me something(s)? Bye, Love, Lindi".

"Dear Mom, I hope you have nice days in Ireland. I wish you didn't have to go. I will miss you. I love you. P. S. I don't want you to die on the plane. At this moment, I'm crying, Love Lindi".

In contrast, our son Raymond, however, never wrote any notes for my trips. (Could this suggest some inherent difference between boys and girls?) While Raymond was at his first year away from home and in college, he sent me a Mother's Day card – the front of the card was designed with prints: *"It's Ultra Woman! - - Super-charged nurturing force! --- kitchen commando! - - - friend, fortuneteller, fashion coordinator! - - - brilliant psychologist! - - - daring adventures of the late 20th Century!"* (A great card designer). Inside the card, he wrote, *"Dear Mom, this card seems to be made for you. Of course it lacks the business side which you have distinguished yourself in, but then, that only applies to one mother in the world. I am doing well in school, and living well. I miss you and home very much. I look very much forward towards returning home. Happy Mother's Day, Love Ray."*

My husband and I raised two wonderful children, and did that while taking care of our jobs and professional commitments. We have given time and energy the best way we know, then, by example and influence. There is nothing more fulfilling than that, however I must say it was not without frustrating moments.

Today, our son, Raymond, earned both a B.S. and M.S. in computer science and electrical engineering from Massachusetts Institute of Technology, and then entered the Harvard University's Medical School (Fig. 5.17). Now he is working on his residency at Mass General, Boston. Our daughter, Rosalind, earned a degree in business economics from Wellesley College (Fig. 5.18), worked with UBS Warburg on investment banking and now is working at Saks Fifth Avenue Headquarters in New York City. More importantly, they both are level-headed happy young people with much promise.

Now the children are in their adulthood. I always enjoy and am touched by their cards for Mother's Day. On a recent Mother's Day, Raymond's card was written: *"Dear Mom:....I thought this card was very fitting...Although I think I recognized you as a woman of grace and wisdom long ago. As I graduate from medical school and start real life, I can't keep but think about everything I had to go through to get here. But all of those things were facilitated by your love and hard work. I don't think anyone else can claim to have had as many opportunities as the ones that you have provided for Lindi and I. Certainly, no one can be as proud of their mother. We are all constantly amazed by how much you have accomplished and your ability to adapt and evolve. Thank you for everything we are today. We love you...Love, Ray".* One of Rosalind's cards: *" Dear Mom:...This card describes perfectly the type of mother you are — loving, unselfish and thoughtful. I know that on numerous occasions you have put Ray and I first and are always looking out for our best interests and happiness...Thank you. Love, Lindi".* For that and for their appreciation, nothing can be more rewarding to a mother.

To have both family and a career and to do well in both is the most demanding task. It takes a tremendous effort and planning. Is it feasible to manage both work and family? We would like to say yes. Excerpts from my speech at the American Association of University Women: *"...* The success of work and family lies in a few considerations: We have to use our full strength, to work with our partners, spouse, and family members to coordinate and to cooperate; think positively, with a can-do attitude. Equally important, we should hold pragmatic views and expectations toward our professors, employers, society, and the country..." (Fig. 5.19)

For the more recent years, I am pleased to serve as "mentor" for younger women who have come to me searching for answers and perhaps comfort as well (Fig. 5.20). In one of my engagements, the president of the Society of Women Engineers (SWE)—Cleveland wrote after an event: *"Dear Dr. Hwang: I would like you to know how much I appreciate your illuminating speech at the SWE on....Because of this successful event, SWE plans for expanding our mentor program....As a Materials Science & Engineering major, I was inspired by your journey to success. I plan to attend graduate school at....Most importantly, your*

Fig. 5.19   Fig. 5.20

Fig. 5.19 Keynote speaker at the American Association of University Women (Cleveland).
Fig. 5.20 Keynote speech at the American Society of Women Engineers—The meeting theme focused on encouraging girls pursuing science and engineering

Fig. 5.21 Hwang looked on President George Bush keynote at the Women's Entrepreneurship in the 21st Century (2004)

story of balancing professional accomplishments with a rewarding personal and family life encourages me to pursue all of my dreams…"

5.2.5 *Thoughts on opportunity*

Opportunity is vital to a person's growth and development, in childhood or adulthood, as a person or in career advancement.

Often, it is that opportunity that brings out the best out of a person.

I would like to share one of my personal experience in the role of mother. When our son was in seventh grade, this was the time that the school is supposed to separate regular classes from the "advanced placement" class in major disciplines, such as math and sciences. For the math class, Raymond was assigned to a regular class. He, at his young age, mentioned to us that he really felt he should be in the advanced placement. I asked him the reasons. (Some kids may just be glad, since it was well recognized that this advanced algebra class requires much homework and study.) He expressed that he can do it and wants to do it. Instantly, it triggered me that this "smart and active" kid must have the opportunity to study hard and work to his best ability. I called the School and then personally visited the Director of the School to communicate Raymond's wish and my thoughts. The Director was courteous and we had a great exchange. Actually the Director had to put up with my preaching about education. But the bottom line position he held was that everything was assigned and the advanced placement class was over-sized already, thus the change could not be made. His reaction was understandable and was even predicted.

But I did not stop there, since I had made the determination that the kid must have the **opportunity** to learn and to develop his talents, especially since the kid did not mind to study harder. I pursued further, adamantly and persistently. Finally the Director made the concession and arranged for me to "negotiate" with the teacher who will actually teach the advanced math class. I did. The teacher was reasonable but not without conditions. The conditions were that Raymond must be willing to go through the whole textbook on Algebra II during the entire summer (3 months), submit the homework every week, take tests and meet with the teacher every other week. In addition, Raymond must, at the end of the summer, take the final exam; and based on his performance and the test score, the teacher will write the report on his work and make a recommendation.

We accepted all conditions. Actually Raymond accepted all requirements.

Not only was I myself impressed by Raymond's acceptance, I was fascinated by it, because this means that he had to give up many of his "plays" during the entire summer, and there was no guarantee either. At his age, I thought that was a tremendous commitment on his part.

Here is the end game. "The teacher's Final Report on Ray Hwang's Summer Work"

"Ray Hwang spent this summer working through the pre-algebra II text, Mathematics: A Human Endeavor, by Harold Jacobs. I sent him the tests for each unit and much of the supplemental work I had done with my pre-algebra II class. Dr. Hwang mailed me all the work Ray did (the first two exercise sets for each lesson, supplemental worksheets and the unit tests). I checked the work as I received it.

At the end of the summer, Ray worked through a supplemental unit on solid geometry. I visited him on August 21st. We covered the necessary algebra vocabulary to algebra I. On August 24th Ray and I met at Hawken to complete the pre-algebra II course. Ray took final exam. We discussed many of the topics he had studied during the summer.

I am satisfied with Ray's work In fact, I am very impressed by his self-motivation and determination. He worked every day to complete a three-trimester course in three months. Although he missed out on the classroom interaction, which is an integral part of Jacob's course, Ray read all the material and completed the exercise sets. His daily work was thoroughly and accurately done. Most sets were 100% correct. This indicates to me that Ray worked very hard. The material was all new to him, and he was able to get a handle on it nonetheless. This is quite a feat!!

I would like to thank Dr. Hwang for supporting Ray in this endeavor. She undoubtedly kept Ray on the right track and helped him get over the rough spots. Without some support, this summer course could not have succeeded. I am grateful for her knowledge and assistance.

Ray is ready for algebra I. I applaud his summer efforts and initiative. Bravo! On a job well done."

Raymond got into the algebra advanced placement class. Since then, he has taken all advanced placement classes throughout his high school years and was at the top of all classes. (Ray was good in sports, arts and other areas as well.) We sent a letter commending this grand math teacher for her dedication to education.

To this day, I know that the opportunity of higher demands that our son's desire to meet and then to fulfill at his young age paved his elevated standing all the way through his schooling to college and then to the medical school, all at his very first choice.

I deeply believe in the importance of opportunity for one's development and advancement.

5.2.6 Reading, writing and speaking

As a ritual, I have made it a point to read or sometimes at least scan over ALL magazines, journals and papers that I have assigned to myself as my reading group. My fixed reading group includes The Wall Street Journal, Business Week, Newsweek, Corporate Board Member magazine and a couple of women's magazines, such as Vogue and W. Then of course about 20 technical or industry-related publications are part of my fixed reading. Every Sunday morning, the flight time in the airplane and the 30-minute every weekday morning of The Wall Street are my routine reading times. Initially, it was a discipline. Gradually it has become a ritual and a relishing experience.

Writing takes time. Writing well takes more time. But writing is a powerful exercise for discipline, for organizing and for brain firepower.

In writing a book, I always appreciated the McGraw-Hill Professional's Guide for Authors. Excerpts:

"It has been said that there are two great flashes of inspiration in the process of publishing a book, the first coming when the project is conceived, and the second when you finally hold a finished copy of the book in your hands. These are separated by a long, seemingly interminable time of perspiration. At first the sweat is almost entirely yours. Later it is ours as well, and it is almost entirely ours in the final stages, after you read the typeset galleys or page proofs.

But the book begins with you. You—the author—are the creator. We—the editors, production coordinators, typesetters or desktop operators, designers, layout and paste-up artists, marketers, salespeople, and others who collectively constitute McGraw-Hill—are the mechanics. Your editor can help you shape the book in its early stages, and he or she can help you visualize the audience for whom you are writing, but the expertise and the writing itself must come from you.

Your book has an excellent chance of success—otherwise we would not have entered this partnership—but both partners will have to contribute an enormous amount of time to secure that success. The venture is not without risk, but the rewards can be substantial....Now, while the light from the first great flash of inspiration is still strong, is the best time to roll up your sleeves and prepare for the perspiration ahead. If you begin now and treat these guidelines seriously, your hard work will be leavened by a growing sense of satisfaction and achievement. Besides, the sooner you submit your manuscript, the sooner you can watch us sweat...."

Indeed, writing a book is an inspiration as well as a perspiration. I recall my writing that first book. It was overwhelming—taking much of my personal time on weekends, evenings and holidays. And it was a learning experience and a great mental challenge. For the subsequent books, the experience learned ranging from organizing the information to handling logistics has facilitated the work. As the old saying goes, the more you do it, the better you are at it.

In technical books, I have just reached my sixth sole-author book. They are:

"Solder Paste: Technology and Applications for Surface Mount, Hybrid Circuits, and IC Component Manufacturing", (ISBN-0-442-20754-9) Van Nostrand Reinhold, New York,1989

"Japanese--Solder Paste: Technology and Applications for Surface Mount, Hybrid Circuits, and IC Component Manufacturing", (ISBN-0-442-20754-9), Nihon, Japan,1991

"IC Ball Grid Array & Fine Pitch Peripheral Interconnections", (ISBN-0-90-115029-0), Electrochemical Publications, LTD, Great Britain, 1995

"Modern Solder Technology for Competitive Electronics Manufacturing", (ISBN-0-07-031749-3, McGraw-Hill, New York, 1996

"*Environment-Friendly Electronics—Lead Free Technology*" (ISBN: 0 901 150 401), Electrochemical Publications, LTD, Great Britain, 2000 "Implementing Lead-free Electronics: A Manufacturing Guide" (ISBN: 0-07-144374-6), McGraw-Hill, 2005

Outside the technical field, I also dabble on issues related to education, business, workforce, and to the impact of technology. Topics addressed include *"Leadership", "Preparation for New Millennium—Education, Technology and Workforce", "Education in Science & Engineering", "Modern Manufacturing Workforce", "Asia's Road to Economic Recovery", "Accelerated Tax Depreciation for High-tech Manufacturing", "Virtual Corporation", "Modern Woman", " Women in Education, Technology and Workforce", "Building Trade and Relations with China", "International Trade and Trade Promotion Authority", "Affirmative Action—Principles & Practices", "Outsourcing or Not and To What Extent", "Changes and Coping with Changes", "Welcoming the Digital Economy", "Globalization—Technology, Trade, and Jobs"...*

During writing, two feelings impressed upon me the most. First, having the ideas and the knowledge stored in your mind is quite different from transferring those ideas and information into black and white words. I also realized a distinction between sole authorship and co-authorship. A sole author of a comprehensive book deals with an overwhelming volume of information that demands exceptional brain power, brain capacity and physical endurance.

Since my first book was released in 1988, I have had the increased opportunities to engage in public speaking in the format of lecture series as well as featured/keynote speeches. I have made a point to allocate some time in "speaking". On average, I receive 20 to 30 invitations for speeches and lectures per year from all over the world. It has been a rewarding commitment.

To sum up what I have learned about public speaking, it comes to three main points: first, the more you do, the better you get; secondly, the preparation for a speech or lecture, albeit time-consuming, is a great way to crystallize one's thoughts and distill one's knowledge on a topic; thirdly, the diverse questions from the audiences are the most intellectually invigorating. And I have enjoyed it all.

5.2.7 Thoughts on globalization

I personally have been fully benefited by the global exposure in business and culture as well as in science and technology over the years.

Globalization is mind-boggling; the more the subject is examined, the more its complexity and intricacy are revealed. Opposing views about globalization and how it relates to technology, job and trade are abundant as reflected through manifold data, survey and debate. Punch lines with various spins include "exports equal jobs", "imports kill jobs", "imports are good for consumers", "offshore R&D is no good", "outsourcing is an integral part of a global trading system", and more... But one thing is clear. We are indeed facing a new world, characterized by uncertainty, changes, choices, flexibility and opportunities.

New economy or knowledge economy will continue being fueled by the ever-swift information flow, ever-fast new knowledge generation, and the way the information and knowledge are used. This new economy is subtly striding into the unfamiliar territory of uncertainties, flexibility, and choices, consequently impacting on business models, and the working environment, as well as the educational system.

During the half century since World War II, the transition from basic agriculture to advanced manufacturing has transformed some countries, as vividly reflected by the industrialization miracles of Korea and Taiwan. U.S. corporations' off-shore operations (manufacturing) have been developing, Ireland in Europe and Singapore in Asia being exemplary in 1970s and 1980s.

Broadly, new and renewed issues in trade and politics, which impact on the globalization of knowledge and jobs, continue to evolve and to be resolved. Business queries continue to be raised and answered. Nonetheless, the Asian region has turned into a hub covering various industries in manifold functions. For example, its electronics and microelectronics industry is journeying through its historically phenomenal record, albeit some countries in their first decade of development, some in their second decade, and some already in the third decade. Despite the political ups and downs between the U.S. and the region over the time, the electronics industry relentlessly marches on. Demands for better and lower cost information-hardware are ever increasing, making the region one of the most vibrant manufacturing areas on the global map.

Global dynamics is such that, in order to maintain the momentum of productivity, three key elements---technology, education, and the workforce---must synchronize with the market place and global thrust.

Several phenomena are becoming obvious for the recent time: R&D (research and development) cycle is shortening; product development cycle is faster in response to market demand; product development cycle is shorter in response to global competition; product life cycle is shortening; manufacturing sector is consolidating; production is being globalized.

To produce more with fewer people and lower cost is becoming the operation's on-going goal, making the ever-increasing productivity a relentless target. This target stimulates profound changes in the shift of job market in nature, geography and the sheer number of jobs on the global scale. For a given function, the productivity level has sharply risen and the number of employees required to perform an equivalent function is much lower than 10 years ago. This productivity and competitiveness-driven environment on the global scale intensify the outsourcing and off-shoring, which have been evolving for many years. Yet the effects and results on jobs only became obvious more recently.

Overall, the global competitiveness drives the customer-centric operation and strategy, and the global competition makes customers at the end of the supply chain dominate the marketplace.

With multiple potential benefits, out-sourcing and off-shoring will continue to grow. It is always a strategic decision to take advantage of the benefit of outsourcing without losing the fundamental knowledge. The critical thought-process goes to the assessment of the core competencies and the sorting out of functions or products for outsourcing from those that should be in-house. With outsourcing and offshoring, the development, protection and ownership of intellectual property are certainly an on-going issue. Nonetheless, the benefits of the ability to utilize the global resources including talents and workforce outweigh the risks.

The U.S. is the largest exporter, as well as the largest consumer in the world. As foreign incomes grow and with the lower price of U.S-produced goods and services, the demand for U.S. exports would expand. In this politically sensitive time, should the free trade be hidden? History tells us that open market and free trade are prerequisite for global competitiveness. The challenge to the private sector and to the government's policy-making is how to formulate a virtuous circle to feed the dynamic economy.

Going forward, all-sourcing (out and in) and all-shoring (on- and off) will become plausible strategies. Consequently, we all have to face these issues. In the long run, innovation and competitiveness are key to a constantly rejuvenating economy. Only a strong economy retains and creates jobs.

5.2.8 Thoughts on future workforce

The future workforce will deal with a new set of challenges under the global competitive environment. The workforce will feature with diversity, ambiguity, an increasing level of complexity, and fast change. Multidisciplinary practice and cross-functional partnership will be the essential part of the working environment. Technology will play an increasingly important role in every industry sector. It is estimated that sixty percent of new jobs in the next two decades will require technological skills.

On an individual basis, skills and knowledge in handling both technical and business demands set the criteria for one's performance and success or lack thereof. With the ever-shortening product cycle and time-to-market, the ability to adapt and to learn is no longer an attribute but a necessity. It becomes a living fact that what one earns depends on what one knows and learns. Simply put, this translates into the level of contribution one can truly make in a timely fashion under the fast-changing and increasingly competitive marketplace.

Collectively, the effort is directed toward nurturing an informed workforce through attaining and developing the following attributes: solid formal education, continued training, life-long learning ability, cost-consciousness, capability to learn and adapt, capacity to grasp new technologies, participation in business process, global views, and internet effects.

From the business (employer) side, it is imperative that the workforce is able to understand how and why a decision is made and a process works. It is also important to be able to follow procedures in the meantime to be able to innovate. Inevitably companies are required to provide timely training to existing and new workers. To run a successful business, cost control (cost reduction) is an on-going effort. Every level of workforce should be more closely linked to "cost". A cost-conscious workforce, as a whole, accounts for the effectiveness and efficiency of an operation.

After the initial evolution for the last decade, the Internet will continue having unprecedented impact on the workforce. It allows team members in different geographical locations and time zones to coordinate efforts, generates more flexibility, provides more current information, and makes timeless and borderless communication possible. One can therefore build on others' ideas and generate solutions on a more timely basis. On the other hand, while one may consider Internet a "disruptive technology", it changes the way to do business and conduct scientific research and product development. It will continue disrupting the traditional balance between supply and demand or the power between buyers and sellers, and streamlining inventories. Prompt adaptation to this disruptive force separates those who are poised to thrive from those who wither.

As the workforce is globalized, U.S. corporations have more access to the large pool of workforce, rendering the ability to leverage a global pool of the intellectual capital.

When a job is to be filled, the company would ask and should ask: would the job be the best fit for the American scientist/engineer for solutions, or would it be better to recruit a foreign worker, or to outsource the job?

The scientists/engineers would ask themselves, what is my value, what do I bring to the job? Am I replaceable? Am I needed?

5.2.9 Thoughts on engineering education

Although I have not been a full-time professor in a university, I have committed to professional advancement teaching all along. This experience has kept me interfacing with the professionals who have an engineering education, which helps me better relate the workforce to higher education.

On the front of engineering talents and pipeline of talents, undergraduate degrees in engineering granted annually in Japan, China, India, and Russia are up to about 103,000, 195,000, 129,000, 82,000, respectively (U.S. Census Bureau). And the U.S. graduates about 60,900 engineering students per year.

Job loss is painful, and in the meantime, the pipeline issue is a legitimate concern. How to reconcile these two fronts?

In a straightforward food chain, better education provides an individual more opportunities in the job market. Collectively it forms a better-educated workforce for the nation, which in turn is the foundation to be globally competitive.

To achieve better education for the nation as a whole, setting high academic standards in core curriculum, particularly Math and Sciences is regarded as the first step. Basic science and math courses form the foundation crucial to any discipline and should not be substituted. Universities under the guidance of and in collaboration with the government must provide all students an understanding of the relevance of science and engineering to everyday life or society. We also need to build in the curriculum the necessary program to facilitate successful transition from school to work through integration with and understanding of industry and market demands.

An expected change in an engineering education is the need to meet the increasing demands in versatility and diversity of skills and knowledge for the graduates to muster their future jobs and career development. The school curriculum needs to prepare the students for their professional specialty. Equally important is nurturing the students with diverse intellectual acumen above and beyond the specialty. The curricular offerings in specialty courses aimed at equipping the students with diverse intellectual acumen above and beyond their specialization will become more important, paving the groundwork for life-long ability to acquire new knowledge, thus retaining jobs.

Some engineering colleges have launched programs that are geared to provide the students with a science and engineering education while fostering the ability to cope with real-world demands. Only those school systems offering the curricula that are integrated with industry interaction and in sync with the market needs are the providers of a premium education. For example, the programs are designed to give the science and engineering graduates a mix of skills in addition to the technical specialty (skills such as ability to deal with customers/clients, handle contracts, help close the deal, work with others….). Programs range from coaching teamwork and tackling manufacturing problems to building written and verbal communication capability and business negotiation acumen. The ability to adapt and to change is another orientation pivotal for success in the future workplace on a global intellectual platform.

Despite impressive advancement, the workforce of engineering is still not as diverse as desired. To quote one female engineer: "The field is still unfriendly to women. The old boy club is alive and well. Regardless of the job a woman does, she stands a high chance of being passed over for rewards and recognition."

Some believe that diversity is one of the greatest things that has happened in this country and within one corporation. Diversity indeed facilitates the creation of new markets and enhances adaptability to the continuously changing market. A diverse workplace enriches the quality of intellectual discourse and brings diverse thoughts and ideas. Managing a diverse workforce to leverage all talents and engage all employees becomes a prerequisite to maximizing productivity, thus competitiveness.

I am all for open and inclusive practice in any level from classroom to board room.

5.2.10 *Thoughts on practical knowledge and entrepreneurship*

I have been challenged to work in both corporate and entrepreneurial worlds. An entrepreneur faces a different set of challenges. Notably, I quickly found out that the corporate "umbrella" that may have sheltered some from "rain and storm" is no longer in existence. It can be a "scary" environment if one is not prepared for it.

The entrepreneurship is considered a primary attribute of enabling the science to move technology to commercialization swiftly.

There is an increased need for the academic researchers to not only possess scientific and engineering expertise, but also the understanding of the market needs above and beyond the science and engineering. To move the scientific knowledge and discoveries from the laboratory to the manufacturing know-how to the marketplace, practical knowledge and entrepreneurial spirit are becoming the niche skills.

Overall, entrepreneurship constitutes multi-faceted attributes, including being able to learn, working hard, being intensively driven, being innovative, being flexible, being versatile, being passionate, being willing to take risks, being adaptive to changes, being super-connected, internet-pace, being technologically savvy, and having the ability to lead.

Broadly, entrepreneurship has been and continues to be an important element to the U.S. economic growth. It also plays a major role to the fast track of industrialization in other countries. Future business models for corporations will have to ignite entrepreneurship within in order to be competitive in the global market (Fig. 5.21).

5.2.11 Thoughts on leadership

I view that leadership applies to everything we do. For example, in the task of chairing a meeting, the leadership inspires discussion and ideas and meanwhile presides over the meeting effectively for its time management and outcome.

I strive to constantly learn about the subtle side of leadership, which I believe is a crucial part of being a leader. It was a joy to receive a note from a meeting attendee, excerpt: " *Dear Dr. Hwang:.....Thank you so much for sharing your evening and thoughts with us last week at "Exchange Horizon". I think everyone at our table was inspired by the time you took to learn about each of us and go beyond the typical introduction...It was lively, inclusive and informative. You did a good job of making sure no one left or dominated the conversation, keeping everyone in the conversation..."*

Leadership has a great impact on the outcome from the War room to the Boardroom. Some believe that leadership is innate, and some believe it can be trained.

Various definitions have been placed on leadership. For instance, General H. Norman Schwarzkopf describes the leadership as the ability to get people to do willingly what they normally would not do. He passionately calls that "character" is the single ingredient most vital to genuine leadership—thus, a leader may not be loved, but must be respected. He offered some rules about leadership: Rule 1—No organization will ever get better unless the leadership recognizes something is broken and needs to be fixed (improved); Rule 2—As a goal is targeted, a leader makes his/her people understand the goal and the role each person plays toward achieving that goal; Rule 3—A leader demands high standard and reinforces success within the organization. The General also passed on two "secrets" for the 21st century leaders--when placed in command, take charge and do what is right.

Indeed, it is not easy to assign a concise and uniform definition of leadership, yet I view that the modern leadership enliven a number of characteristics: confident, intellectual rigor, innovative thought-process, having diverse exposures and versatile knowledge, ability to ask the right questions, ability to motivate and inspire others, penchant for intense dialogue, social-consciousness, relation-building, ability to weave a clear vision, willing to set high standards, in the position to demand high standards, ability to influence, and being respected (may not be liked).

I also hold the view that leadership can be learned and nurtured. Let us always learn to be a better leader.

5.2.12 Pioneering and long-standing work in SMT

What drives the end-use market are the continued convergence of computing, communication and entertainment as well as the relentless growth of the wireless, portable, handheld digital electronics and optoelectronics. On the ever-changing technology landscape, the industry has responded and will continue to respond to competitive demands in the global marketplace. New electronic gadgets will be featured with increasingly higher functionality, further simplicity, lower cost and greater operational ease. What has transpired from these market demands is continued technological innovation and an ever-shortening product life cycle.

Electronics miniaturization is phenomenal. Surface Mount Technology (SMT) continues to be the backbone in manufacturing ever-changing electronics.

Surface Mount Technology (SMT), as the name implies, is the application of science and engineering principles to the design and manufacturing of electronic circuitry by placing components and devices on the surface of circuit boards in lieu of the traditional through-hole connection. The SMT, in simple terms, provides superior performance/cost ratio for manufacturing printed circuit board, which is the "brain" of most modern products.

Encompassing the advent of the electronics age 25 years ago to the current day, I have been dedicated to the establishment of the infrastructure of surface mount technology in electronics manufacturing. The effects of my work, behind the scene, can be found in virtually all walks of life, including consumer and household electronics, communications, computers, entertainment, automotive, industrial, defense, oil drilling, aerospace and more. Being a part of the solution in terms of manufacturing and production problems and failures, has been highly rewarding to me. I have visited and worked with almost all circuit board manufacturers around the world during the most exciting and dynamic electronics age, in a role of a material supplier as well as of an advisor to the industry.

While I was serving as the national president of the Surface Mount Technology Association, our team was able to expand the national organization to abroad, including the formation of several international chapters and creating the global network for the industry.

Surface mount has been a critical manufacturing technology for electronics PCB assembly for more than two decades and is expected to play an important role in the foreseeable future. Consequently, solder paste will continue to serve as the most viable interconnecting material for electronics assembly mass production.

Solder paste technology is a true multidisciplinary and interdisciplinary practice.

5.2.13 Leadership in environment-friendly lead-free electronics

In the electronics industry, the market demand and an ever-shortening life cycle of electronic gadgets have been driving the technological development. In the meantime, environment-friendly manufacturing and the delivery of environmentally benign end-use products that are ultimately safe at the end of the product life cycle will become essential to technology-business competitiveness. This is a continuing challenge to the industry.

The overall goal is to design a product for minimal environmental impact and with the full lifecycle in mind by eliminating the use of highly toxic substances. In this case, it is to eliminate the use of lead in electronics. It should be noted that using environment-friendly materials and processes are not intended to replace recycling. Continued effort on developing economic recycle technologies in conjunction with starting with environment-friendly materials/processes is the sound practice in the long run.

The ultimate goal of producers and manufacturers is to build a sustainable world.

Today, achieving environment- friendly electronics has become a global movement, analogous to the CFC movement fifteen years with this global objective.

5.2.13.1 *Development of lead-free electronics*

The concerted effort in research and development of replacing lead in electronics applications started in the late 1980s. One pioneering effort was initiated under the U.S. Defense Mantech Program and U.S. Army Materiel Command with the goal of reducing the cost and enhancing the reliability of electronic weapons on a national scale by improving the scientific basis of process controls used on production lines, focusing on the soldering technology thrust area to improve and advance material properties. The goal was also to enable the use of surface mount electronics assemblies in severe thermo-mechanical environments, as well as to develop the manufacturing system needed for an environmentally friendly lead-free matrix. I was invited to participate as an advisor for the program. The United States was the pioneer, working actively on the issue of lead elimination and reduction in the first half of the 1990s in both technological development and legislation proposals. Other development programs, individually and in consortia, have been pursued during the 1990s.

As a result of sustained and systematic research in conjunction with continued research and development work in the private sector, the technology has indeed advanced and the foundation has been paved.

My book *"Environment-Friendly Electronics: Lead-Free Technology"* (ISBN: 0-901-150-401) covering lead-free technology in preparing for implementation and the book "Implementing Lead-Free Electronics—A Manufacturing Guide" (ISBN: 0-07-144374-6) focusing on the actual implementation and production are the culmination of the fifteen years steadfast research and development work. These two books are designed to be complementary to each other in providing the broad-based information from fundamental technology to production to reliability.

Steve Chapman, Editorial Director of McGraw-Hill Professional commented on the book: *"...I would like express the pride that McGraw-Hill takes in publishing Dr. Jennie Hwang's new book. The challenges of lead-free manufacturing are many, and will affect the electronics industry for many years to come. This book is an implementation roadmap and an invaluable reference tool to those charged with making these new techniques work on the factory floor. As such, it takes sophisticated theory and applies it in a manner both clear and eminently practical. This is McGraw-Hill's mission – the clear transmission of technical information from the realms of the theoretical into the workplace, with the goal of improving businesses, products, and lives – and nowhere in our publishing program this year is this mission better expressed than in Dr. Hwang's landmark work."*

5.2.13.2 *What made the commercial success?*

During the course of development, many have argued that the lead-free technology had no purpose in the science and technology world and that the lead-free product had no place in the market.

Nonetheless, our team was not distracted and has continued to steadfastly keep our eyes on target. Today, lead-free electronics is a necessity and is a global commitment. Our unwavering and unrelenting effort has made us not only a leader in technology, but also in commercialization.

5.2.14 *Prime of life – into the future*

Time to slow down? No way!

Looking into the future, the most inspiring figure is Richard W. Pogue, Esq. Mr. Pogue, as the Managing Partner of Jones Day for many years, is not only one of the most successful lawyers having helped build one top law firm, but also a phenomenal leader in various capacities from technology community to higher education arena. After more than four decades of professional services and leadership, Dick at age of 73 is still as vibrant as ever, going to office six days a week and continuing to make contributions to legal, industry and educational sectors. I mentioned to Dick: "…aspiring to your dedication, I definitely will have another solid twenty (20) years to contribute…"

My sight is far to the future, to new technology, to new business, to new horizons to explore and contribute. I view the next twenty years as the prime of my life.

Life has been intensive and enriching personally and professionally. With the children grown and working on their own careers, perhaps a little time to smell the roses (as some friends urge me to do, and even children often "tell" me so) (Fig. 5.22). I have resumed the singing and dancing that I was fond of in my youth but was unable to do for the past 25 years—I discovered I can still hold a microphone and sing (Fig. 5.23). I love ballroom dancing and fashion, and today can still make steps (Fig. 5.24). I am equally passionate about business strategy, investment and selected world issues.

Going forward, in the role of mother, instilling the ethic of learning and hard work into the children will continue to be on my agenda. Among all, I am most proud of these two children and they hold so much promise (Fig. 5.25).

Fig. 5.22   Fig. 5.23

Fig. 5.24

Fig. 5.22 Have a bit fun time in Milan, Italy (2004)
Fig 5.23 Resume a bit of singing at leisure (2003)
Fig 5.24 Ballroom dance with husband (2004)

Fig. 5.25   Fig. 5.26

Fig. 5.25 Mother-daughter quality time (2004)
Fig. 2.26 Conferring Dr. Jennie Hwang Award to a college senior, Amanda Yoho at YWCA.

In support of science and engineering, a YWCA Award recognizing outstanding women students who study in science or engineering was established (Fig. 5.26). I have also established an endowment at Case Western Reserve University, designated to encourage faculty and students to acquire international exposure, which I view as increasingly important to the higher education. Additionally, a Faculty Excellence Award was set at Cleveland State University, honoring faculty's exceptional research and service performance.

Having opportunities to interact with a diverse group of people in both technology and business arenas across national borders has provided me not only with professional gratification, but also with insight into the subtleties of people and business dealings. Two important experiences I want to share are that a solid business relationship must be a win-win deal and a real business success has to cover the immediate gain as well as the long-term leverage. I strive to use that intricate understanding in my future work.

Writing and speaking have been wonderful forums for me, serving as a nice perch from which to share a global view. It is a demanding discipline, but when I receive welcoming and appreciative feedback, it is a most rewarding and joyful feeling. I continue to share my thoughts through writing and speaking on issues, particularly those about which the future global businesses are concerned.

Regardless of much has been learned, nothing can replace the hard work. I want to continue to work hard and contribute the most.

Not until recent years, have I felt the power of confluence of versatile and diverse exposures and experiences that I have worked for and built. This powerful confluence has been most utilized in my corporate board service, which is also intellectually appealing. I will continue to allocate a portion of my time to do that.

After being tested and having gone through close scrutiny, I feel I can do almost anything. I want to open up the new facets of my career. While pursuing what are in place wholeheartedly, I look to the opportunity of exercising true leadership requiring intellectual rigor and critical and strategic thinking.

Chapter 6

Douglas D. Osheroff: Winner of the 1996 Nobel Prize in Physics

6.1 Introduction by the Editor

6.1.1 *The Nobel Prize*

The Nobel Prize, founded by Alfred Nobel, is the first international award given yearly since 1901 for achievements in physics, chemistry, physiology or medicine, literature and peace. In 1968, the Prize given in memory of Alfred Nobel in economic sciences was instituted. Each prize consists of a medal, personal diploma and prize amount. Winners of the Nobel Prize in physics include Joseph J. Thomson of United Kingdom for his theoretical and experimental investigations of the conduction of electricity by gases (1906), Albert Einstein of Germany and Switzerland for his discovery of the law of the photoelectric effect (1921), and Niels Bohr of Denmark for his investigation of the structure of atoms and of the radiation emanating from them (1922). In 1996, the Nobel Prize in physics was awarded jointly to Douglas D. Osheroff (Stanford University), Robert C. Richardson (Cornell University) and David M. Lee (Cornell University), all of USA, for their discovery of superfluidity in helium-3.

6.1.2 *What is superfluidity in helium-3?*

Helium is the second most common element in the universe. The most common element is hydrogen. Helium mostly exists in natural gas from the interior of the earth, although it also exists in the atmosphere of the earth. There are two forms (called isotopes) of natural helium, namely helium-4 (with two protons and two neutrons in the nucleus of the atom) and helium-3 (with two protons and one neutron in the nucleus of the atom). Helium-4 accounts for 99.999% of natural helium, whereas helium-3 accounts for only 0.001% of natural helium. However, helium-3 can be produced in appreciable amounts in nuclear reactions.

Superfluidity refers to the phenomenon in which a fluid flows without friction. Thus, a superfluid cannot be kept in an open vessel; the fluid creeps as a thin film up the vessel wall and over the rim. Kammerligh Onnes was the first to produce a superfluid in 1909, but he did not note this result. Pyotr Kapitza, who won the Nobel Prize in physics in 1978, was able to show that the viscosity of liquid helium-4 below its lambda transition (phase change associated with the shape of the specific heat versus temperature curve resembling the Greek letter λ) is 0, and hence, it must be a true superfluid. This result was observed to occur at 2.17 K (K means Kelvin, with 273 K = 1°C). Superfluidity was expected in helium-4, but not in helium-3, due to the difference in statistical law for these two types of atoms. However, superfluidity is possible in helium-3, due to the pairing of the atoms, thereby allowing helium-3 to obey the same statistical law as helium-4.

6.1.3 *Scientific contributions of Dr. Osheroff*

The scientific contributions of Dr. Osheroff go beyond the discovery of superfluidity in helium-3. Although materials are usually used at ordinary temperatures such as room temperature, fundamental physics involving electrons or small particles are most effectively studied at low temperatures (temperatures below a few degrees K). This is because the thermal excitation of the electrons or atoms at temperatures that are not sufficiently low causes difficulty in observing certain phenomena associated with the electrons or atoms. Through extensive investigation of phenomena at low temperatures, using experimental techniques such as magnetic, electrical, optical and thermal measurements, Dr. Osheroff has advanced fundamental understanding of how fluids and solids behave at low temperatures.

6.1.4 *Honors received by Dr. Osheroff*

In addition to receiving the 1996 Nobel Prize in physics, Dr. Osheroff received the Oliver E. Buckley Condensed Matter Physics Prize (1981), the Sir Francis Simon Memorial Award (1976), the MacArthur Prize Fellow Award (1981), the J.C. Jackson and C.J. Wood Chair in Physics (1992), and the Richtmyer Memorial Lecture Award from American Association of Physics Teachers (1998). Furthermore, Dr. Osheroff is Honorary Senator of Heidelberg University (2002), Foreign Associate of Korean Academy of Science and Technology (2000), and Member of U.S. National Academy of Sciences (1987).

Fig. 6.1 Dr. Osheroff

6.1.5 *Career development of Dr. Osheroff*

Dr. Osheroff was born and raised in Aberdeen, Washington, USA. He received his B.S. degree in Physics from California Institute of Technology in 1967, and his Ph.D. degree in Physics from Cornell University in 1973. In 1972-81, he was Member of Technical Staff of AT&T Bell Laboratories. In 1981-87, he was Head of Solid State and Low Temperature Physics Research in AT&T Bell Laboratories. Since 1987, he has been Professor of Physics and Applied Physics in Stanford University.

In addition to being an outstanding researcher, Dr. Osheroff is an effective administrator. He served as Chair of the Department of Physics in Stanford University in 1993-1996 and in 2001-2004.

6.2 Dr. Osheroff's Description of His Life Experience

Ethnically, I come from a mixed family. My father was the son of Jewish immigrants who left Russia shortly after the turn of the century, and my mother was the daughter of a Lutheran minister whose parents were from what is now Slovakia. Mostly, however, I grew up in a medical family. My father's father and all his children either became physicians or married them. My parents had met in New York where my father was a medical intern and my mother was a nurse. At the end of the Second World War, my parents settled in Aberdeen, a small logging town on the west coast of Washington State, where medical doctors were in short supply. Surrounded by natural beauty, it was a perfect place to raise a family, and I was the second of five children.

To this day I grow pale at the sight of blood, and never for a moment considered a career in medicine. Despite this, my father, who was usually engrossed in his medical career, inspired in me passions for both photography and gardening, which were his hobbies when time permitted, as they are mine. Natural science interested me intensely from a very early age. When I was six I began tearing my toys apart to play with the electric motors. From then on, my free hours were occupied by a myriad of mechanical, chemical and electrical projects, culminating in the construction of a 100 keV x-ray machine my senior year in high school.

My projects often involved an element of danger, but my parents never seemed too concerned, nor did they inhibit me. Once a muzzle loading rifle I had built went off in the house, putting a hole through two walls. On another occasion a make-shift acetylene 'miners' lamp blew up on my chemistry bench in the basement, embedding shards of glass in the side of my face, narrowly missing my right eye. With blood running down my face, I came up the stairs cupping my hands to keep the blood off the carpet. My mother was by then at the top of the stairs. Knowing my propensity for practical jokes, she exclaimed loudly "If you're kidding I'll kill you!" As usual, my father lectured me about safety as he sewed the larger wounds closed, and there was always an unspoken understanding that that particular phase of my experimentation was over.

In high school I was a good student, but only really excelled in physics and chemistry classes. While I liked physics much more than chemistry, the chemistry teacher, William Hock (Fig. 6.4), had spent quite a bit of time telling us what physical research was all about (as

Fig. 6.2 Osheroff family. Douglas Osheroff stands to the far left of the children.

Fig. 6.3 Douglas Osheroff, seated in the middle, playing cards.

Fig. 6.4 Science club in high school. Douglas Osheroff is standing just above and to
the left of the chemistry teacher, Mr. Hock (wearing a lab coat).

opposed to my experimentation), and that effort made a deep impression
on my young mind. My interest in experimentation helped me to
develop excellent technical skills, but I did not feel motivated to do
independent reading in those areas of physics or chemistry associated
with my projects. I was intellectually rather lazy, and in high school I
would always take one free class period so that I could get my homework
out of the way, freeing the evenings for my many projects.

William Hock, my chemistry instructor in high school, left Aberdeen,
Washington, just a couple of years after I was his student in order to
teach in Los Angeles. While I was an undergraduate student at Caltech,
he brought a group of his high school students there for a tour of Caltech

and we met again at that time. In 1981, when I received one of the first 21 MacArthur Prized Fellowships, I thought about those people who had influenced my career in science and his name came to mind very quickly. I decided to see if I could find his phone number, assuming that he was still in Los Angeles and was eventually able to give him a call. When he answered I simply said: "Mr. Hock. Guess who!" Without hesitation, and not having seen or heard from me in some 16 years, he replied instantly: "Why, that must be Douglas Osheroff." Of course he by then knew that I had received the MacArthur Award, but I was still astounded. I also contacted him after the announcement of the Nobel Prize, and we once got together when I was down in Los Angeles at a Stanford Alumni Meeting. I last sent him an e-mail (or called, I cannot remember which) in March 2005. I was attending an American Physical Society Meeting in Los Angeles, but he was busy and we could not get together.

My parents were generous, and the home for me was filled with scientific toys and gadgets. In addition, their children were allowed to attend any university to which we could get admitted. I chose Caltech over Stanford to avoid a continuing comparison of my academic record with that of my older brother, then a Stanford undergraduate.

It was a good time to be at Caltech, as Richard Feynman was teaching his famous undergraduate course. This two-year sequence was an extremely important part of my education. Although I cannot say that I understood it all, I think it contributed most to the development of my physical intuition. The Feynman problem sets were very challenging, but I had the good fortune to know Ernest Ma, who was an undergraduate one year ahead of me. Ernest would never tell me how to solve problems, but would give obscure hints when I got stuck, at least they seemed obscure to me at the time.

It was a shock to suddenly have to work so hard in my studies. I had the most trouble in math, and only through considerable trauma did I gradually improve my performance from a grade of C+ to A+ over a three year period. Years later, when Caltech was offering me a faculty position, I confided that I did not have a very illustrious career as an undergraduate. To this remark the division chair replied "That's OK, Doug. We are not hiring you to be an undergraduate."

The pressure at Caltech was extreme, and I am not sure I would have survived had I not joined a group of undergraduates working with Professor Gerry Neugebauer on his famous infrared star survey during my junior year. This experience made me recognize how satisfying research could be, and how different it was from doing endless problem sets. In my senior year, in order to get out of a third term of senior physics lab, I also began working in David Goodstein's low temperature lab (David was in Italy). Two professors, Don McCullum from University of California at Riverside and Walter Ogier from Pamona College, were spending their sabbatical leaves there trying to reach a temperature of 0.5K by pumping on a helium bath in which the superfluid film had been carefully controlled. They filled my mind with the wonders of the low temperature world, and I decided I would go into solid state physics.

I chose to attend Cornell for graduate school largely because it was so far away from the Pasadena smog. In the end, it was a good choice, and a good time to be at Cornell. Soon after my arrival I met two people who were to become very important in my life. While still looking for housing, I met Phyllis Liu, a pretty young woman from Taiwan, who had also just arrived in Ithaca. We dated a bit, but then she found herself too busy with her studies for such diversions. We met again three years later, and were married in August, 1970, two weeks after she obtained her Ph.D. The other person was Professor David M. Lee, the head of the low temperature laboratory at Cornell and the professor under whom I was to work as a teaching assistant my first year. Dave seemed to think that I was bright, and encouraged me to join the low temperature group.

Low temperature physics seemed even more exciting at Cornell than it had been at Caltech. New technologies and interesting physics made the field easy to choose, and I found myself thoroughly enjoying every minute of my work. In the spring of my fourth year, Professor David M. Lee (Fig. 6.5, with whom I later shared the 1996 Nobel Prize in Physics) asked me to talk to the Bell Labs recruiter, who came to campus in the fall and spring of each year. I was not ready to graduate, but we talked a bit, especially about making tiny electrical plugs to be used throughout the Bell Telephone system. It seemed interesting to me, although not really physics. In the fall, Dave suggested I start interviewing in earnest.

Fig. 6.5 David M. Lee (1931-), who shared the 1996 Nobel Prize in Physics with Douglas D. Osheroff (1945-)

Fig. 6.6 Dr. Osheroff in the laboratory during the discovery of superfluidity in ^{3}He.

I first talked with General Electric, who seemed to have no jobs whatsoever. I then talked to Bell Labs again, but this time to a new recruiter, Venky Narayanamurti, who had recently received his Ph.D. in physics at Cornell. Venky was enthusiastic about what I was doing, and felt that I might be able to get a postdoc doing Raman spectroscopy. I didn't confess that I knew nothing about the subject.

We discovered our mysterious phase transitions in my Pomeranchuk cell in November 1971, and almost by magic, Venky called me up in early December with good news. The hiring freeze which had been in place for almost two years at Bell had been lifted. How soon could I be ready to come down for a job interview? I told Venky that we had stumbled on to something that was pretty exciting, and we fixed the date: January 6, 1972. Fig. 6.6 shows a photograph that I took of myself in the laboratory during the discovery of superfluidity in ^{3}He.

At Bell Labs, a job interview began with a thesis defense, and it could at times turn nasty. I was lucky that no one questioned my association of the A and B features with the solid. In particular, Dick Werthamer was in the audience, and he had done early work on the p-wave BCS state soon to be associated with the B phase. I think my enthusiasm carried the day, and ultimately Bell Labs offered me not a postdoc position in Raman spectroscopy, but a permanent position which would allow me to continue my studies on superfluid in ^{3}He.

Phyllis and I moved to New Jersey in September, 1972; Phyllis to a postdoctoral position at Princeton University, and I to Bell Laboratories in Murray Hill. This was the golden era at Bell Labs. The importance of the transistor, invented in the research area there, made management extremely supportive of basic research. The only requirement was that work done should be 'good physics' in that it changed the way we thought about nature in some important way. I joined the Department of Solid State and Low Temperature Research under the direction of C. C. Grimes, and began purchasing the equipment I would need to continue what I by then knew were studies of superfluidity in ^{3}He. Some instrumentation was even purchased before I arrived in New Jersey. Yet I knew it would take at least a year to set up my laboratory, and I feared that most of the important pioneering work would be done before my own lab became operational.

I was surprised to find that by the time my laboratory did become operational, few of the studies that interested me had been done. Indeed, there seemed to be some question as to whether or not these new phases were all p-wave BCS states. In addition, theorists Phil Anderson and Bill Brinkman at Bell Labs had become interested in the theory of superfluid ^{3}He. This set the stage for what was to be an extremely productive period in my career. Over a five year period, beginning in 1973, we measured many of the important characteristics of the superfluid phases which helped identify the microscopic states involved. We found the superfluid phases to be almost unbelievably complex, and at the same time extremely well described by the BCS theory and extensions to that theory developed during that period.

In about 1977 I began to feel pressure from Bell Laboratories management to go on to study other physical systems. I decided to study solid ^{3}He, my original thesis topic, and at the same time Gerry Dolan and I began a modest program to test some of the ideas that David Thouless had discussed on electron localization in disordered one dimensional systems. This latter study had to fit within the extremely slow time scale of the solid ^{3}He work. By late 1979, both of these efforts had succeeded beyond my wildest expectations. We discovered antiferromagnet resonance in nuclear spin ordered solid ^{3}He samples which we grew from the superfluid phase directly into the spin-ordered solid phase. At the same time, the low temperature group at the University of Florida also discovered these resonances, but because we cooled our samples by adiabatic nuclear demagnetization of copper rather than Pomeranchuk cooling, only we were able to form and study single crystals, and could thus identify the allowed magnetic domain orientations. In the end, Mike Cross, Daniel Fisher and I were able to determine the symmetry of the magnetic sublattice structure, and correctly guessed the precise ordered structure, later confirmed by polarized neutron scattering. The frequency shifts resulting from this antiferromagnetic resonance have made solid ^{3}He an extremely useful model magnetic system, and to understand them theoretically, we had borrowed some of the same formalism which Leggett used to understand the frequency shifts in superfluid ^{3}He.

At almost the same time that Cross, Fisher and I made our breakthrough in our solid ^{3}He studies, Dolan and I discovered the log(T) temperature dependence to the electrical resistivity in disordered 2D conductors which Phil Anderson and his 'gang of four' had just predicted would exist, as a result of what they termed 'weak localization'. I did not continue the work on weak localization, as I only had one cryostat, and to do so would have meant that I could not continue my studies on nuclear spin ordering in solid ^{3}He, since the two sets of experiments would have vastly different time scales. Somewhat ironically, I got a second cryostat two years later.

In 1987, after fifteen years, I left Bell Laboratories to accept a position at Stanford University. I had received informal offers of university positions periodically while at Bell Labs, but always found Bell to be the ideal place to do research. The combination of in-house support for basic science and first rate collaborators made Bell Labs unbeatable as an environment for doing research. However, my wife recognized in me a teacher waiting to be born. In addition, she was not happy with her job in New Jersey, and we agreed that she would apply for positions elsewhere. When she received offers from two biotech companies in California, Amgen and Genentech, I suggested that she accept the Genentech offer and that I would start talking to Stanford and U.C. Berkeley. Stanford, which has a small physics department, had just begun a search for a low temperature physicist. Ultimately, I received offers from both institutions, and chose Stanford because we liked the atmosphere better, and it was a better commute for Phyllis.

At Stanford my students and I have continued work on superfluid and solid ^{3}He, studying how the B superfluid phase is nucleated from the higher temperature A phase and diverse properties of magnetically ordered solid ^{3}He in two and three dimensions. In addition, we have developed a program to study the low temperature properties of amorphous solids. Our work has shown that interactions between active defects in these systems create a hole in the density of states vs. local field, just as is seen in spin-glasses. In amorphous materials, it may be possible to measure the size of coupled clusters of such defects, something which has been difficult in spin-glasses.

I have thoroughly enjoyed all aspects of university life, except for having to apply for research support. In particular, I have been fortunate to have had excellent graduate students, and to be able to teach bright undergraduates. Of course, with undergraduates one always has a few students who do not appreciate the professor's efforts. In 1988, after teaching my first large lecture course, one student wrote in his course evaluation: "Osheroff is a typical example of some lunkhead from industry who Stanford University hires for his expertise in some random field." Despite this minority opinion, in 1991 Stanford presented me their Gores Award for excellence in teaching. From 1993-1996 I served as Physics Department chair, and stepped down in September 1996, hoping to spend more time with my graduate students. The day I learned I was to receive the Nobel Prize, after just two and a half hours sleep the night before, I taught my class on the physics of photography, although the lecture was not on photographic lenses, but the discovery of superfluidity in ^{3}He.

Chapter 7

Klaus Biemann: The Father of Organic Mass Spectrometry

7.1 Introduction by the Editor

7.1.1 *What is organic mass spectrometry?*

Other than the bones, teeth and water in various parts of your body, your body is essentially a collection of organic substances, such as proteins, fats and carbohydrates. Organic substances are also central to plants, animals, food, pharmaceuticals, contact lenses, fabrics, tires, plastics, electrical insulators and gasoline.

In general, an organic substance contains carbon as the key element. The carbon atoms are commonly linked to form a chain, which serves as the backbone of a molecule. Atoms such as oxygen, hydrogen and nitrogen are bound to the carbon atoms in various parts of the chain. Side chains can also be linked to various parts of the main chain. In an alternate geometry, the carbon atoms are joined to form a ring. Because of the large variety of geometries for linking the various atoms and the enormous number of atoms that can be in a single molecule, the number of types of organic molecules is unlimited. Furthermore, an organic substance material, such as blood, typically consists of numerous types of molecules. Therefore, the synthesis, analysis and use of organic substances constitute a rich field of science and technology.

The analyses of organic substances include the identification of various types of molecules in the substance, the determination of their concentrations, and the finding of the type and arrangement of atoms in each molecule. Without such analyses, science would become a black art. However, with pertinent analysis, one can understand the precursors and the properties of each substance. Then one is able to design substances for specific applications by tailoring the molecules involved. Thus, organic analysis is critical to the development of pharmaceutical agents and numerous other products that are of immense significance.

Mass spectrometry is the most powerful technique of organic chemical analysis. It is not a wet chemical method carried out in a test tube, but involves the use of an instrument called a mass spectrometer. This instrument functions by ionizing molecules, breaking up a large molecule into fragments, and separating the ionic fragments by the use of a magnetic field, an electric field or other means. The masses of the fragments are then measured to assemble the original structure.

7.1.2 Scientific contributions of Dr. Biemann

Dr. Biemann is a chemist who has been extremely successful and versatile in applying modern methods of organic chemical analysis to the study of a wide range of biologically and technologically important molecules. In particular, he is a pioneer in the use of mass spectrometry for the elucidation of organic substances. Dr. Biemann is noted for the successful application of this technique to the analysis of proteins, carbohydrates, drugs, metabolites, fossil fuel combustion products, natural products, and the chemistry prevailing on the planet Mars.

In addition to being an outstanding researcher, Dr. Biemann has been a dedicated professor. His 40 years of chemistry teaching in both the graduate and undergraduate levels have nurtured generations of chemists that are active professionally in industry, research and education all over the world. In particular, Dr. Biemann has nurtured a total of about 140 graduate research students and postdoctoral fellows. In addition, Dr. Biemann has written a widely acclaimed textbook on mass spectrometry and has over 350 publications in scientific research journals.

7.1.3 *Honors received by Dr. Biemann*

The honors received by Dr. Biemann are too numerous to be listed in full. He is a member of the U.S. National Academy of Sciences (elected in 1993). He received the Award in Analytical Chemistry from the American Chemical Society (2001), the Pehr Edman Award for Outstanding Achievements in Mass Spectrometry (1992), the Thomson Medal (1991), the Newcomb-Cleveland Prize from the American Association for the Advancement of Science (1978), the Fritz Pregl Medal from the Austrian Microchemical Society (1977), the Exceptional Scientific Achievement Award from the U.S. National Aeronautics and Space Administration (1977), the Powers Award from the American Academy of Pharmaceutical Sciences (1973), and the Stas Medal from the Belgian Chemical Society (1962).

7.1.4 *Career development of Dr. Biemann*

The scientific career of Dr. Biemann spans over five decades and reflects his immense dedication to research and teaching, his diligence, versatility and adaptability in research, and his solid training in chemistry.

Dr. Biemann was born in Austria in 1926. He is married to Vera and blessed with two grown children, Hans-Peter and Betsy. He received his Ph.D. degree in organic chemistry from the University of Innsbruck, Austria, in 1951. He moved to the USA in 1955 and became a USA citizen in 1965. In the Department of Chemistry of the Massachusetts Institute of Technology (Cambridge, Massachusetts, USA), he was Research Associate in 1955-1957, Instructor in 1957-1959, Assistant Professor in 1959-1962, Associate Professor in 1962-1963 and Professor in 1963-1996. Since his retirement as Professor Emeritus in 1996, Dr. Biemann has been living in New Hampshire, USA.

7.2 Dr. Biemann's Description of His Life Experience

The last day of World War II found me in a small village north of Dresden (Germany). A friend and I decided to avoid, at any cost and risk, falling into the hands of the Soviet Army. At that night we took off on bicycles with a loaf of bread, towards the southwest. During the next night we abandoned the bicycles and crossed the river Elbe just downstream from Dresden in a small boat and continued on foot towards the mountains and what had just become again Czechoslovakia. That region had been occupied by American forces a few days earlier but was at that time being turned over to the Russians in accordance with the agreement the Allies had signed just three months earlier at Yalta (Crimea, USSR). So we quickly turned west to get over the line into the region controlled by the Western Allies. On the fourth day of our hike we bode each other good-bye, my friend going north while I turned south, hoping to reach a tiny village near Innsbruck (Austria) where my mother with my two older sisters had sought refuge before Vienna was occupied by the Soviet Army. This was a 600 km walk, along country roads, sometimes through woods and fields to by-pass occasional checkpoints. These were set up, because in these days, immediately after the end of the war, people were not allowed to move far beyond their communities. There was nothing to eat, except an occasional raw egg or a boiled, but cold, potato obtained from a kind, understanding farmer or his wife. But that was the least of my concerns.

It happened to be Mother's Day when I reached the village after a two-week hike. My family was indeed there, in the home of my mother's friend who had offered a place to stay in those turbulent times.

7.2.1 *Student years*

My late father had been a pharmacist, and so was my oldest sister. Thus it was understood that I would also study pharmacy. The University of Innsbruck was only about 25 km away, but there was no transportation, neither train nor bus, for several months. So one sunny summer day I just walked there (and back) to see whether I could enroll for the fall term. Fortunately, my mother had taken my high school certificate with her, so I had no difficulty registering. However, housing was a problem, because not only were student dormitories virtually unknown at Austrian universities, but many apartment buildings in the city had been heavily damaged by bombing during the later part of the war. But I was lucky: my mother's cousin and his family, who lived in Innsbruck, offered to let me stay with them during my studies.

The students entering in the fall of 1945 were an odd assembly of ages and stages in their academic life. In fact, I was the only male student entering his first semester in pharmacy or chemistry (the first four terms of both had about the same curriculum). There were a few older men, who had to interrupt their studies for the draft and had returned, but most of their classmates, if still alive, were prisoners of war somewhere, a fate I had successfully avoided by my little walk. Women had been able to continue and complete their studies. Therefore, there were only those few who just had finished high school and were ready to enter the university.

In contrast to the American college system, with its rigid four-year curriculum, pace is less structured, even leisurely, on the continent. There it is up to the student to accumulate all the courses, exams and – at least in chemistry and pharmacy – laboratory requirements needed to obtain the final degree. Because of the aforementioned composition of the student body, I could immediately obtain a space in the laboratory and thus get ahead of the large number of students who entered in the later semesters. But in that first winter there was one problem in carrying out chemical experiments in a speedy manner: because of the severe shortage of gas, as well as any other fuel, the city turned on the gas supply lines only for two hours during the day. So our professor told us to bring two bricks with us, heat them over the Bunsen burner and then stand on them to keep warm while carrying out our experiments. Since most of them also required gas, the days in the laboratory were short.

With time conditions began to normalize, but life was frugal and detractions were few. So I had nothing better to do but to pursue my program and was soon ahead of all my contemporaries. When I finished my studies in pharmacy after two and a half years with the obligatory "Magister" (master's) degree, I realized that I really preferred to be an organic chemist. This was quite easy, because – as mentioned earlier – the first few semesters of the curriculum in both fields were very similar and I just needed to continue with the more advanced chemistry courses.

In the additional laboratory program I was again the only student in my category, the older ones finished and the new ones behind. The newly appointed Head of the organic chemistry department, Professor Hermann Bretschneider, assigned me to try out a new, more modern set of laboratory experiments. After that, I became his first graduate student, an easy choice for him, because there was no one else, and *vice versa.* Three years later, in February of 1951, I graduated with a Ph.D. degree in organic chemistry.

I relate the story of this period of my life, because it had taught me to take risks in making decisions that led me off the well trodden path and going it alone, if necessary. So after a few years as an instructor in organic and pharmaceutical chemistry in Bretschneider's department, I became a little restless. By chance, in the fall of 1953, I noticed on the bulletin board of the Dean's office a little announcement about a summer program at the Massachusetts Institute of Technology (MIT) in Cambridge, USA. To make a long story short, in spite of Bretschneider's tacit disapproval, I applied, was accepted and boarded the *S.S. Maasdam* of the Holland-America Lines from Rotterdam to Boston (*via* New York) the next May.

I really had no idea where I was going until someone on the boat asked me what I will do upon arrival. When I said that I have a Fulbright fellowship to spend the summer at MIT, they almost fell off their chairs. This gave me the first inkling that the Institute must be something special. Soon I learned about the hierarchy and pecking order of American Universities and that MIT was near – some people would say at – the top.

The program I was to attend was called the "Foreign Student Summer Project" (FSSP). It had been initiated a few years earlier by a group of MIT undergraduate students, who had interrupted their studies (voluntarily or through the draft) to spend the last year in the Armed Forces during the war in Europe. They had witnessed first hand the destruction of many cities and knew that most of the universities had suffered greatly. Returning to MIT to finish their studies, a group of them decided to help. They established a program that allowed about 60 young scientists from all over Europe, and later the entire affected world, to come to MIT and use its facilities to carry out research that could not be done at their home universities. It was for those who had completed their studies and held positions for the academic year, but could get away for the summer. The program was supported by the Alfred P. Sloan Foundation. (Mr. Sloan was an alumnus of MIT and President of General Motors Corp.).

I was assigned to the laboratory of George Buechi, an assistant professor who had been appointed recently after a postdoctoral stint at the University of Chicago. The Head of MIT's Chemistry Department, Professor Arthur C. Cope, felt that it would be easier for me to work with someone who also spoke German. This was considerate and reassuring, but not really necessary. I had five years of English in high school, which was mainly devoted to reading, writing and translating. So my grammar and spelling were quite good and I now had an opportunity to perfect my English speaking skills. The ability to express myself well in written and spoken English was very important for my later academic career. Buehi was Swiss and had earned his Ph.D. in organic chemistry at the Eidgenoessische Technische Hochschule (ETH) in Zurich (the MIT of Switzerland). His work involved the determination of the molecular structure and synthesis of natural products, i.e., substances produced by plants or animals.

This brief stay at MIT, which I was able to extend to the end of November, greatly expanded my horizon. To the thorough training in synthetic organic chemistry I had obtained during the years with Bretschneider, now had been added knowledge of the use of physical methods, such as ultraviolet and infrared spectroscopy, which had not yet reached Innsbruck. Similarly, I became better acquainted with modern mechanisms of organic chemical reactions based on "electron pushing"[1]. Such approaches had not been widely used by the professors teaching my generation. They ignored them, considering the new concept as "not invented here" and relying instead on memorized experimental facts and experience. The other thing I learned to know was an entirely different

---

[1] The concept that for a chemical reaction to occur electrons have to move in concert, breaking bonds between atoms and simultaneously making new bonds between others.

academic system, very much in contrast to the autocratic hierarchy dominant in Austria (and Germany as well).

When I arrived back in Innsbruck in early December of 1954, I dramatically felt this difference. Professor Bretschneider treated my absence like a black hole, never asking me what I did – or even learned – during these six months abroad. I resumed my teaching obligations in the classroom and in the organic chemistry laboratory. The rest of the time was spent working on Bretschneider's research projects. The next step in my career would be the "Habilitation", a prerequisite for further advancement. This process required the demonstration of independent, significant research. After a few months I brought up this topic. Professor Bretschneider told me that, in due time, I would be permitted to publish some of the work under my own name to fulfill that requirement and later could work on my own research projects, whenever I am not needed to help out on his. I was not very enthusiastic about this plan, because I knew that such help would always be needed. Consequently, I decided to go my own way again.

7.2.2 *Moving to MIT*

To make a long story short, when George Buechi learned that I did not want to stay in Innsbruck, he offered me a position as postdoctoral research associate[2] and by the beginning of October 1955 I was back in his laboratory at MIT.

After the Second World War, when research at universities had returned to normalcy, many of the best organic chemists focused their attention on the determination of the structure of natural products and their synthesis. The intellectual challenge was to learn the chemical make-up of substances produced by microorganisms, plants, animals or even humans. Many of these were used as medicines (like penicillin, morphine etc.), fragrances (menthol, camphor etc.) or had other important biological functions (steroids, vitamins etc.). Once a plausible

---

[2] A person, usually at a university, who has obtained a doctoral degree, but wishes to receive additional professional training. "postdoc" for short.

structure of the molecule had been deduced on the basis of chemical reactions and physical properties, the ultimate proof of its correctness was the synthesis of the molecule from known starting materials by a series of well understood chemical reaction steps.   An important ingredient for success in this field was a good understanding of the way nature performs chemical reactions ("biogenetic considerations") and of the detailed mechanism by which chemical reactions proceed in the laboratory.   In both areas research groups in the US, the UK and Switzerland were predominant, a position previously held by Germany, but lost first by the emigration of many Jewish scientists beginning in 1933 and then the physical and intellectual devastation during World War II.

The research group of Professor Leopold Ruzicka (Nobel Prize 1939) at ETH in Zurich (Switzerland) was a leader in the field of "terpenoids" (compounds related to turpentine) to which many flavors and fragrances produced by plants and animals belong. George Buechi had obtained his Ph.D. degree there and had retained not only the interest in that field, but also the connections of Ruzicka's group to Firmenich & Cie. of Geneva (Switzerland), an internationally known company producing fragrances for the perfume industry and flavors for food companies. Firmenich supported part of Buechi's research at MIT, including the position I was holding.

One of the projects I worked on was the synthesis of a compound the Swiss group had isolated in 1946 from the perfume gland of the musk deer. They named it "muscopyridine". The highly priced natural musk had by then been replaced by a chemically unrelated compound with the same smell, so the synthesis had no commercial value. It was undertaken to prove a structure (I) of the compound (see Appendix A), which Buechi had derived on paper based on the hunch ("biogenetic consideration") that it may be made by the musk deer from another compound, called muscone, which is also present in the animal's perfume gland. Starting with a commercially available substance, a graduate student had successfully carried out the first five steps of the synthesis (sufficient to get his Ph.D. degree), but had run out of material before he could complete the next six. So I had to repeat his work and then finish it.

Any synthesis of a substance that is to prove a proposed molecular structure has to be very rigorous if it is to serve this purpose beyond any reasonable doubt. The product of each of the 11 individual chemical reaction steps had to be characterized. This required careful purification followed by measuring the compound's elemental composition. For that, a small amount (a few milligrams) of it has to be burned to carbon dioxide, water and nitrogen and these products carefully measured. Such analyses are performed by specialists to whom one submits a sample and waits (and pays) for the results.

After a few months of careful work I reached the last step one day and it was late at night that I was ready for the crucial test, the "mixed melting point"[3] of my synthetic product with a sample of natural muscopyridine. The mixture of the two melted indeed at the same temperature as the individual compounds: that means that the two were identical and the elaborate synthesis proved that the proposed structure (I) was indeed correct!

7.2.3 *Starting the academic career*

Hard work paid off. Before I was seriously considering the next move in my career, Professor Cope asked me whether I would be interested in joining the faculty as Instructor (at that time the first step of the academic ladder at MIT) in the Analytical Chemistry Division of the department. An organic chemist himself, Cope felt that this sub-discipline represents 80% of all chemistry and should, therefore, be represented in Analytical Chemistry by someone trained in organic chemistry. Apparently he had been quite impressed by my synthesis of muscopyridine. He also knew from my curriculum vitae that I had taught a course in the analysis of pharmaceuticals at the University of Innsbruck and thus appeared to be a logical choice for that new faculty position, eliminating the need for a nationwide search. That I had just

---

[3] Identity of two compounds of the same melting point is established by mixing them together and checking whether the mixture still melts at the same temperature (identity) or lower (non-identity). This principle can be best illustrated with water: it freezes (and melts) at $0°$ C; if it is mixed with more water, it still freezes at $0°$ C; but if it is mixed with something different (like salt or antifreeze) it freezes at a much lower temperature.

married an American girl, Vera, seemed an additional indication that I was committed to serious life in the U.S. Although many mainstream organic chemists would not have moved into the analytical direction, I saw it as an opportunity and accepted Cope's offer. As of July 1, 1957, I was a junior member of the faculty at MIT.

However, having an appointment in the Division of Analytical Chemistry caused a slight problem: I had to choose a new field of research outside of the main stream of organic chemistry in which I had been trained all the years in Innsbruck and then polished at MIT. So I thought of ways by which I could utilize my knowledge in organic reactions and structure determination to turn it into problem solving in the qualitative or quantitative analysis of organic compounds. The projects also had to be suitable for obtaining research grants.

One idea was quite ambitious and interesting. Towards the end of my years in Innsbruck, I had worked out a chemical reaction which converted an organic acid like acetic acid to a very stable five membered ring consisting of two carbon and three nitrogen atoms, called a "triazole". While that work had been aimed at making compounds that may become pharmaceuticals, I now thought of an entirely different use: the determination of the amino acid sequence of peptides and, ultimately, of proteins.

Proteins are one of the most important substances in living systems. They perform, regulate and control all chemical reactions in the cell. Proteins consist of a string of as few as 20 to more than 1,000 amino acids linked together by "peptide bonds". Since there are twenty different amino acids[4], there is an almost unlimited number of proteins possible which all differ in the sequential arrangement ("sequence") of these amino acids. Certain parts of such sequences are designed by nature to perform a particular process (e.g. hemoglobin carries oxygen), while other portions may differ from species to species (cow or human, for example). Even small changes can have dramatic effects: the change of a single amino acid in human hemoglobin causes the red blood cells to

---

[4] All amino acids have the general structure $H_2N-CHR-CO-OH$, but differ in the structure of the group R. The peptide bond (-CO-NH-) is formed by the connection of one amino acid to the next through elimination of water (H_2O) between the -CO-OH group of one and the H_2N- group of the next (see structure II).

become distorted and inefficient, causing sickle cell anemia, a debilitating, inheritable disease!

It had been barely five years earlier, that Professor Frederick Sanger (Cambridge University, UK) had, for the first time, determined the structure (the arrangement of amino acids in the molecule) of a protein, insulin. This was accomplished by separating the molecule into two chains 21 and 30 amino acids, respectively, in length and then chopping each chain into smaller pieces by treatment with acid. This produced mixtures of mainly individual amino acids, dipeptides (two amino acids still connected), tripeptides (three amino acids) and small amounts of longer peptides. The free amino acids do not, of course, tell which others they were connected to. But for dipeptides (A-B) Sanger had developed a method that did: by attaching a chemical marker (*) at one end (the "amino group") of the peptide (*A-B) followed by acid cleavage ("hydrolysis"). Identification of the marked amino acid *A revealed that the dipeptide was A-B, not B-A. In a tripeptide A-B-C, one could tell that A was at the amino end but could not distinguish A-B-C from A-C-B and had to look for a dipeptide B-C or C-B, to go further. Sanger received the Nobel Prize for this work in 1958.

If one also could mark (') the other ("carboxyl") end of a tripeptide, it would be possible to establish the sequence of a tripeptide in a single, two-step experiment that would produce *A-B-C'. Even a tetrapeptide would leave only two possibilities, *A-B-C-D' and *A-C-B-D', instead of the six possibilities if only A is marked. So I planned to use the triazole formation I had worked out in Innsbruck to mark the carboxyl group of peptides.

7.2.4 *Learning about mass spectrometry*

Such ideas, if successfully executed, might have made me a reasonable analytical chemist, were it not for a seemingly unrelated event. While I was still working in Buechi's group, Firmenich wanted to know what was going on at a flavor and fragrances conference held in Chicago in the spring of 1957. Rather than sending someone over from Geneva, they asked me to attend and provide a report. I gladly agreed, not so much because I was interested in the talks (there was not much of my kind of organic chemistry going on) but more because I could take my first ride on an airplane! The only talk I still can remember was by W. H. Stahl, from the Quartermaster Corps Laboratory in Natick, Massachusetts (not far from Cambridge), who spoke about the identification of flavors from various fruits. They were all small molecules, like acetone, methyl butyrate, butyl acetate, etc., as analyzed by "mass spectrometry". I had never heard of this method, which apparently was widely used in the petroleum industry for quantitative analysis of crude oils and gasoline. Stahl showed that he could reliably identify these small, very volatile molecules by matching their fragmentation pattern ("mass spectrum") with that of known compounds.

A simple example illustrates the basic features of a mass spectrum. When ethanol (ethyl alcohol) molecules are bombarded with an electron beam at very low pressure (a vacuum), they loose an electron to form positively charged molecular ions, $CH_3CH_2OH^+$. Sufficient energy is left in these ions that in some of them a bond breaks, leaving a positively charged fragment ion. When these ions are subjected to an electric and a magnetic field, they separate according to their mass. When the electric field is decreased or the magnetic field increased, the ions pass across a detector and are recorded, one after the other, by increasing mass. The detector measures the abundance (number) of ions at each mass and a recorder plots the data as shown in Figure 7.1. In the formula the bonds that are broken to lead to the peaks in the spectrum are indicated by dotted lines; the numbers correspond to the mass of the resulting fragments (sum of atomic weights, $H = 1$, $C = 12$, $O = 16$). The height of each peak is an indication of the ease by which the bond(s) break to form the particular fragment, which in turn is influenced by the chemical structure.

So, when I was later thinking about my future research projects, it dawned on me that one might be able to omit many of these chemical manipulations by simply measuring the mass spectrum of the compound of interest. The mass spectra of the small molecules Stahl was dealing with were obviously related to their structures. I figured that this should also hold for much larger and more complex molecules, and ultimately make it possible to determine the structure of new compounds.

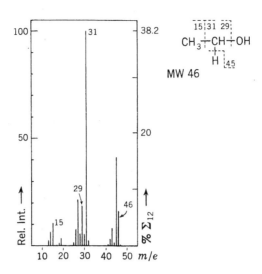

Fig. 7.1 The mass spectrum and chemical structure of ethanol.

A few days in the library revealed a number of papers, some by F. W. McLafferty (then at Dow Chemical Co., Midland, Michigan) describing the correlation of the mass spectra of simple organic compounds with their structural types. Also Professor Einar Stenhagen (at the University of Uppsala, Sweden) was carrying out similar correlations for the esters of fatty acids and related compounds. And, of course, the petroleum industry had assembled large collections of the mass spectra of hydrocarbons.

However, I was not interested in such routine data collection and correlation work with known, commercially available compounds. I planned to determine the amino acid sequence of small peptides derived from naturally occurring proteins of unknown structure. Since there are 20 different amino acids that make up proteins, there are 400 different dipeptides and 8,000 tripeptides. Therefore, assembling a complete "library" of mass spectra of all possible peptides for comparison was out of the question. Instead, one had to interpret the spectrum of the unknown "from scratch".

There was, however, a much more serious problem with my idea: in order to obtain a mass spectrum, the compound in question had to be vaporized into the mass spectrometer. Because peptides have a basic group and an acidic group in the same molecule, they form internal salts, so called "zwitter ions" (II for a tripeptide), which are not volatile. Even when heated in the vacuum of the mass spectrometer's sample inlet system, these molecules just decompose to a char. Therefore, no one had ever attempted to get a mass spectrum of such materials. Here my training and experience in organic chemistry came into play: I knew about a few simple chemical reactions that would convert a peptide into

$$\overset{+}{H_3N}\text{-}\underset{\underset{R_1}{|}}{CH}\text{-}\overset{\overset{O}{||}}{C}\text{-}NH\text{-}\underset{\underset{R_2}{|}}{CH}\text{-}\overset{\overset{O}{||}}{C}\text{-}NH\text{-}\underset{\underset{R_3}{|}}{CH}\text{-}\overset{\overset{O}{||}}{C}\text{-}O^-$$

II

$$CH_2\text{-}CD_2\text{-}NH\text{-}\underset{\underset{R_1}{|}}{CH}\text{-}CD_2\text{-}NH\text{-}\underset{\underset{R_2}{|}}{CH}\text{-}CD_2\text{-}NH\text{-}\underset{\underset{R_3}{|}}{CH}\text{-}CD_2\text{-}OH$$

an analogous molecule (III) which retains all its structural characteristics but would be volatile and thus amenable to mass spectrometry. I also expected that these "derivatives" of peptides would fragment very specifically, revealing the sequence of the amino acids.

The only remaining obstacle was that I did not have a mass spectrometer, an instrument costing more than $50,000, much more than any instrument used by organic chemists at that time. When I asked Art Cope why the department did not have one, he replied "It takes an electrical engineer to keep it running". I said, "I do not think so. I can handle it myself." Cope replied: "If you promise that it will not collect dust, I will come up with the money". We both kept our promises.

I ordered a Consolidated Electrodynamics Corp. model 21-103C mass spectrometer, the instrument then mainly used in the petroleum industry for quantitative hydrocarbon analysis. It was delivered and installed in May of 1958 (Figure 7.2).

Fig. 7.2 The mass spectrometer commercially available in the 1950s.

In the meantime, the grant applications, which I had submitted to the National Institutes of Health (NIH) and the National Science Foundation (NSF) for the chemical peptide sequencing had been approved. Fortunately, the rules permitted changes in approach, if a better way was found. I felt that using mass spectrometry would be allowed. But I now had to build up my own research group to carry out the experiments. It was quite unlikely that I would be able to persuade incoming graduate students to work with a newly appointed faculty member without any track record and on a topic (mass spectrometry) which they had never heard of. To begin with postdoctoral associates was the alternative and I had already included such a position in both the NIH and the NSF proposals. So I wrote to two former students in Bretschneider's laboratory in Innsbruck, who had recently graduated and were working in small pharmaceutical companies in Austria. Both, Josef (Sepp) Seibl and Fritz Gapp, gladly accepted my offer to join me at MIT. Sepp arrived just the day the instrument was delivered. He thus was able to participate in its installation, a good way to learn from the service engineer first hand how it works. Fritz came over just a few weeks later and we were ready to go.

Like myself, Sepp and Fritz were well trained in experimental organic chemistry and they both started with the peptide sequencing work. Because this was a long range and high risk project I myself began to explore the potential of mass spectrometry in a very different area.

7.2.5 *Alkaloids*

While working on the synthesis of muscopyridine I had learned much about other nitrogen containing natural products: alkaloids. The elucidation of the structure of these rather complex basic (i.e. alkaline, like ammonia) substances, which are produced mainly by plants, had been an intellectual challenge to chemists for many decades. Furthermore, many alkaloids have unique pharmacological activities and have been used for centuries as medicines. The painkiller morphine isolated from opium, the stimulants caffeine from coffee beans and cocaine from the coca nut, are well known examples. In their quest to find new drugs, pharmaceutical companies were always looking for new alkaloids by extracting them from plants. The ones showing interesting biological activities were then studied in detail.

In the 1950's the Swiss company CIBA scored a huge financial success with one of the first antihypertensive drugs, the alkaloid reserpine. It had been discovered in 1932, but its complete chemical structure was only determined 22 years later. For all these reasons many academic and industrial laboratories were feverishly working on the structures of the many newly isolated alkaloids. But this work was tedious and slow. One had to convert the natural product painstakingly to compounds of known structure by well understood chemical reactions. In essence, the process is the reverse of a multi-step synthesis, such as the one of muscopyridine described earlier. The crucial endpoint is again the proof of identity of the conversion product with a substance of known structure.

It occurred to me that this task could be greatly simplified by the use of mass spectrometry and I set out to try it on a relatively simple case. I knew from the literature, that three research groups had independently suggested a structure (IV) for the alkaloid sarpagine based on biogenetic considerations. Only one of them promised to also prove it by conversion to a degradation product (V) of another alkaloid of known structure, called ajmaline. However, more than three years had passed without a further word that this had been accomplished.

Sarpagine belongs to the group of "indole alkaloids", all of which contain an aromatic (benzene-like) indole structure to which is attached an intricate polycylic (multi-ring) system, as schematically shown in Figure 7.3. The shaded part represents the aromatic indole system, the open circle the polycyclic portion. I expected the latter to cleave characteristically in the mass spectrometer, while the former remains intact. As a consequence, the mass spectrum would be dominated by the fragmentation of the non-aromatic polycyclic portion along lines *a* and *b*. For a pair of compounds differing only in the mass of the substituent R attached to the indole system, this difference would only result in a

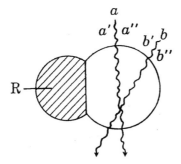

Fig. 7.3 Schematic general depiction of the make-up of an indole alkaloid (see text).

constant, predictable mass shift of fragments *a'* and *b'*, while *a''* and *b''* would be of the same mass in both spectra.

One of the problems with the promised chemical correlation of sarpagine with ajmaline was that the conversion of the former to the known degradation product of the latter was chemically very difficult, time consuming and would require a large amount of sarpagine. The conventional approach for the proof of structure requires *absolute* identity of the compounds to be compared by the "mixed melting point" method.[3] However, if my hunch was correct, mass spectrometry would require only the comparison of two *similar* compounds. The elaborate and difficult conversion of the indole segment of sarpagine[5] would then be unnecessary. The few remaining chemical steps[6] required to make compound VI were simple and could be carried out with the small amount of sarpagine I had been able to obtain. A mass spectrum of the product of each step right away assured me that the reaction was successful and eliminated the need for careful purification and elemental analysis by combustion of part of the precious product.

It so happened that the structure of ajmaline had been determined by Professor Robert B. Woodward at Harvard University, just up the river from MIT. In the course of that work he had made the degradation

---

[5] Removal of an -OH group and addition of a -CH₃ group on the aromatic system.

[6] Cap the -OH group on the aromatic system with a -CH₃ group and remove the -OH group and the double bond from the non-aromatic system.

product V. I knew him very well and he was happy to provide me with a sample. Woodward, the preeminent natural products chemist of that time (perhaps ever? Nobel Prize 1965) was, of course, very curious to learn how my novel approach to structure determination would work out.

The mass spectra (see Figure 7.9, Appendix B) of the two compounds indeed exhibited the exact same fragmentation pattern, with all peaks shifted by the predicted mass difference (net one oxygen atom = 16 mass units). However, for a novel methodology without previous precedent, rigorous demonstration of its validity was necessary. So I obtained a pair of indole alkaloids with a different polycyclic system (X and XI in Appendix B). Their mass spectra also showed an identical fragmentation pattern but it was very different from that of the sarpagine/ajmaline pair. This method of structure correlation of alkaloids (and then also other compound classes) became known as the "mass spectrometric shift technique".

Before this work appeared in print, I presented it at the biennial conference on natural products organized by the International Union of Pure and Applied Chemistry (IUPAC) held in Australia in August of 1960. My talk, generically entitled "Application of Mass Spectrometry to Natural Products" (when submitting the abstract I did not yet know what specific results I would have in time) was scheduled for a relatively small lecture room. But word about my work had gotten around through Woodward and others and the room was packed so that people had to sit on the stairs. Right after the presentation, Professor Carl Djerassi (Stanford University), a well-known steroid chemist and inventor of "The Pill", came over and asked me to spend some time at Stanford to help set up a mass spectrometer in his laboratory and teach his students in its use. For the next two IUPAC conferences, held in Brussels, Belgium (1962) and Kyoto, Japan (1964), I would already be invited as a plenary speaker. This international recognition speeded my promotion to tenured full professor by 1963.

Now the structure determination of alkaloids had become easy. What had previously taken years to accomplish could now be done in a few weeks. With two other postdocs[2] who joined my group from Innsbruck, Gerhard Spiteller and Margot Friedmann, we isolated 16 alkaloids from the bark of a tropical plant, *Aspidosperma quebracho blanco*. Of these only one was of known structure, but its mass spectrum provided enough information to allow us to determine the structure of the remaining 15 within a few months. These structures were deduced by detailed interpretation of the fragmentation processes that led to the observed mass spectra. Here my experience and understanding of the mechanism of chemical reactions played an important role.

Because our methodology was so new, the *Journal of the American Chemical Society* first did not want to publish these results, but wanted us to perform also all the conventional procedures, like elemental analysis by combustion, melting points, etc. This was, however, exactly the point we wished to make, namely that this is no longer necessary. Furthermore, we would have had to isolate much more of some of the compounds, then crystallize them to get a melting point and burn them for the elemental analysis. It took a few explanatory letters to convince the editors that our conclusions are valid. The paper finally appeared in early 1963.

This work on alkaloids and related topics made mass spectrometry almost over night well known to organic chemists. A book I published in 1962, entitled *Mass Spectrometry, Organic Chemical Applications*, was written for that readership and described many of the practical tricks we had developed but could not be included in journal publications. The book became a classic text. Soon every respectable chemistry department in the U.S. wanted to own, or at least have access to such an instrument. For me it meant that students and postdocs were eager to join my research group and grant support became ample. The use of mass spectrometry in organic chemistry, which was more in qualitative rather than quantitative analysis, brought about changes in the market place that led to improved instrumentation more specifically suited for this new field.

In 1962 I obtained funds from NIH and NSF to acquire a new "high-resolution" mass spectrometer, which had a much higher resolving power than my first one. It enabled us to measure masses with an accuracy of one-thousandth of a mass unit. This made it possible to determine the number and kind of the elements present in the molecule and in its

Fig. 7.4 Photograph taken in 1963, showing the author (in back) and Dr. Walter McMurray standing, and Dr. Peter Bommer (seated) in front of the CEC 21-110 high resolution mass spectrometer.

fragments[7]. Such data greatly facilitated the interpretation of the mass spectra of compounds of unknown structure.

An early (1964) and most dramatic result with this methodology was the determination of the structure of the "dimeric" indole alkaloid vinblastine, the first anti-cancer drug (developed and marketed by Eli Lilly Co., Indianapolis, IN), particularly useful for the treatment of Hodgkin's disease, one of the most common types of lymphoma. It had been shown previously that vinblastine was some combination of two known indole alkaloids, velbanamine and vindoline, but it had not been possible to determine its elemental composition reliably by combustion

---

[7] The exact atomic weights of the elements are not round numbers. Based on an arbitrary standard of C = 12.000000, H = 1.007825, N = 14.003074, O = 15.994915, etc. Therefore, measuring the mass of a molecule (or fragment) very accurately makes it possible to calculate how many of these atoms make up the molecule (or fragment).

analysis. The high-resolution mass spectrum gave a molecular weight of 810.4219, which fits only the composition $C_{46}H_{58}N_4O_9$, adding up to 810.4204. From the exact mass of various fragments we could deduce how velbanamine and vindoline were connected to form vinblastine (VII). At the same time we determined the structure of the related alkaloid vincristine (R_2 = CHO instead of CH_3 in VII), which today is still used as part of the drug mixture given to cancer patients during chemotherapy.

$$R_1 = COOH_3 \; ; \; R_2 = CH_3 \; ;$$
$$R_3 = OCH_3 \; ; \; R_4 = COCH_3$$

Over the ensuing years mass spectrometry became indispensable in organic chemistry. It now plays an important role in the chemical and pharmaceutical industry, particularly in drug development. Its chief characteristics - high structural specificity and high sensitivity (it requires less than one millionth of a gram, an almost invisible amount) - make it much faster and cheaper to develop new materials and facilitate quality control.   The need for the wasteful elemental analyses by combustion had practically disappeared.

Drug discovery no longer relies on searching nature for biologically active compounds. Over the last few decades it has been replaced by "rational drug design", a process in which chemists make large numbers (tens of thousands) of compounds they speculate may have medicinal significance. These complex mixtures ("libraries") are then tested for biological activity. A positive result is then followed up by enrichment and purification of such leads. Mass spectrometry plays a very important role in these novel approaches for finding new drugs.

7.2.6 *Nationwide collaborations*

By that time (1965) NIH began to realize that mass spectrometry will play an important role in organic and biochemistry, requiring the training of scientists in this new field. Therefore, the agency asked me to apply for a newly established type of grant to develop a training program for graduate students and postdocs. I pointed out that I would need a second high-resolution mass spectrometer to accommodate the additional people and my own computer to process the large volume of data produced by that instrument. I was told that NIH would award me also one of the new "instrument facility" grants. These were the days of rapidly expanding federal support of research at American universities. I enjoyed this generous grant support continuously for over 30 years.

From that time on, my laboratory had the opportunity and the resources to become involved in many research projects of others, whenever mass spectrometry could make a significant impact. Most often it was planned, but sometimes accidental. Once one of my graduate students attended a cocktail party where he got into a conversation with an anesthesiologist from the Harvard Medical School, whose specialty was life support of comatose patients. When he asked what my student was doing, he was told "determining the structure of alkaloids using gas chromatography and mass spectrometry". The physician was familiar with gas chromatography (see below) but had never heard of mass spectrometry, and soon dropped the subject. Weeks later, he called me to ask for help with a compound he found in the blood of one of his comatose patients. He thought it would be a new metabolite of the drug the man had taken an overdose of, but we identified it as something entirely different and unrelated to the ingested drug.

While disappointed, the anesthesiologist realized the power of our methodology and we developed a program for the rapid detection of drugs in body fluids (blood and urine) by gas chromatography-mass spectrometry. It became especially useful for saving infants and children who were brought to emergency rooms because they had accidentally swallowed a harmful drug. The same methodology is now widely used to detect the taking of illicit drugs by athletes or given to race horses and the like.

7.2.7 *Combining gas chromatography with mass spectrometry (GCMS)*

In the early 1950s A. T. James and P. J. A. Martin in the UK had developed "gas-liquid chromatography" for the fast and facile separation of complex mixtures of organic compounds. It involves passing the mixture in the form of a vapor through a long (2-3 m) tube filled with something similar to sand coated with a kind of grease. The vapor is flushed through the tube by a "carrier gas", usually helium. The components of the mixture are held up by the grease more or less (very simply speaking) in relation to their molecular size and emerge at the other end of the tube one after the other. A detector senses the vapor and a recorder plots a graph representing the amounts and the time it takes for each one to emerge. This graph is a "gas chromatogram".

This method is generally used for the quantitative analysis of mixtures. We used it, however, to get a sample of the components of a mixture to determine its mass spectrum. For that purpose, each emerging "fraction", as indicated by the detector, was carefully collected by sticking a small glass tube over the end of the gas chromatographic tube and cooling it with a small piece of dry ice to condense the sample. The glass tube was then placed into the inlet system of the mass spectrometer and heated to vaporize the sample into the instrument.

This procedure worked fine for our purposes but was tedious and time consuming. Doing it hundreds of times gave all of us ample opportunity to think of ways to eliminate this chore by feeding the gas emerging from the gas chromatograph directly into the mass spectrometer. This simple concept had only one problem: the amount of helium flushing each component was much too much for the pumping system of the mass spectrometer, which would loose vacuum and cease to operate, or perhaps be destroyed! So we had to think of ways to remove the helium before it could enter the spectrometer, but without losing the compound of interest. Finally, one of my graduate students, Jack "Throck" Watson, constructed a simple device which accomplished the purpose. The gas chromatograph was connected to the mass spectrometer by a porous glass tube to the outside of which was attached a vacuum pump. The dimensions of the pores were calculated to efficiently pump helium through the wall but let the large molecules pass into the spectrometer.

This device became known as the "Watson-Biemann separator" to distinguish it from the "Ryhage separator" developed simultaneously by Ragnar Ryhage at the Karolinska Institute, Stockholm (Sweden). His achieved the same result as ours, but based on a different physical principle (jet expansion). The practical difference was the fact that Ryhage patented his design, but we did not. He licensed the patent exclusively to a Swedish instrument company (LKB) and one could only obtain Ryhage's device by purchasing the entire LKB mass spectrometer system. On the other hand, any good glassblower could make our device by simply following the description in our publication (1964). As a consequence, many mass spectrometer manufacturers incorporated it in their instruments, thus making "GCMS" widely available.

The recording of a mass spectrum "on the fly", i.e. during the few seconds when a compound emerges from the gas chromatograph, had been made possible by the availability of mass spectrometers capable of scanning across the entire spectrum in a few seconds, rather than the 20 minutes required by the previous generation of instruments. This trend was driven by the fact that organic and biochemists used the mass spectrometer chiefly for compound identification rather than for quantitative analysis, which had required very accurate peak height measurements.

Once we had begun to use our GCMS system, we realized that it would be much easier to keep the mass spectrometer scanning continuously every few seconds during the entire gas chromatographic separation (15-60 minutes) than to trigger it only when the detector indicates an emerging compound. This is because for a complex mixture some components may overlap. To avoid the piling up of reams of recording paper, we decided to accumulate the spectra directly in a computer. Thanks to the NIH Mass Spectrometry Facility grant mentioned above, I had acquired an IBM 1800 computer in 1966 which was ideally suited for this purpose. With its accessories (tape drives, disks drives, line printers, etc.) it occupied an entire room (Figure 7.5), although its capacity and speed was much less than today's simple laptop. However, for its time it was on the cutting edge and what we needed to develop all the data acquisition, processing and interpretation software required. It served as the basis of the commercial systems used for years to come. Desktop GCMS instruments with built in computers are now widely used, particularly in the analysis of air and water for pollutants.

Fig. 7.5 The IBM 1800 computer in the author's laboratory in 1966. Left: the CPU (operated by Ed Ruiz); right: cabinet holding three disk drives.

7.2.8 *Peptides and proteins*

Returning to 1958 and the peptide sequencing project mentioned earlier, Sepp Seibl and Fritz Gapp had soon worked out the chemical conversion of di- and tripeptides (II) to the corresponding di- and triamino alcohols (III). To our great satisfaction, the mass spectra of the latter were as simple and structure-specific as I had predicted. Since the aim was to ultimately sequence proteins, we had to devise a way for dealing with complex mixtures of small peptides. We decided to use gas chromatography (see the preceding section 7.2.7) to separate the much more volatile amino alcohols derived from a mixture of peptides.

In order to establish that the mass spectra of all possible small peptides (as their amino alcohol derivatives) can be readily and reliably interpreted in terms of their structure, we had to get a representative number of them. In those days few small peptides were commercially available. Therefore, I wrote for samples to a number of biochemists working on the synthesis of peptides or the degradation of proteins. Most of them were kind enough to respond to my request, not only because I needed only very small amounts, but also because they were curious about what I was doing. This had the unexpected side effect, that word got around quickly in the closely knit and small community of people working on the determination of the chemical structure of proteins, a field of great importance, but still in its infancy. As a consequence and even though we had published by 1959 only one short communication (in *The Journal of the American Chemical Society*), I was invited to present a paper at the European Peptide Symposium held in Basle, Switzerland, the following year (Figure 7.6). From that time on peptide biochemists started to pay attention to mass spectrometry.

Demonstrating a principle and applying it to practical cases are, however, two different things. So we spent the next few years on the improvement and refinement of the methodology. It had to work with the very complex mixtures produced by the degradation of a protein using acid or enzymes at the level of a few milligrams of starting material yielding only micrograms of each peptide. Here the high sensitivity of the mass spectrometer was of importance, combined with

Fig. 7.6 Photograph (detail) of the participants in the European Peptide Symposium, August 1960. The author is Nr. 36. The two other Americans are Prof. Klaus Hofmann (64) and Dr. Miklos Bodanszky (11).

the ability of gas chromatography to separate very complex mixtures in a matter of 30 – 60 minutes. The development of our GCMS system described above (Section 7.2.7) greatly facilitated this task. Improvements in the chemistry extended our methodology to hexapeptides, which facilitated protein sequencing.

Other protein sequencing research did, of course, not stand still either. Pehr Edman in Sweden (and later in Australia) developed a chemical method for the stepwise removal and identification of one amino acid after the other from the "amino end" of a protein. Once that approach had been automated and was commercially available, most protein sequencing was done that way. However, there were a number of situations in which it did not work. For example, if the amino end is acylated (chemically blocked), which is the case for many mammalian proteins, the first amino acid cannot be removed and the process cannot start. For our mass spectrometric approach this was no problem but rather a simplification because the first step in our conversion to the more volatile amino alcohols was the acetylation of the amino group anyway (see Appendix C).

Because our method required a certain degree of expertise and expensive instrumentation, it did not lend itself to occasional use in an ordinary biochemical laboratory. Therefore, we were often asked not only to sequence these blocked peptides, but also to determine the structure of the blocking group, which practically "fell out" of the mass spectrum. For certain proteins, particularly those that span cell membranes, a combination of the Edman method and ours was most effective. The determination (1979) of the structure and function of rhodopsin, the protein occurring in the retina and involved in vision is one example.

Soon thereafter it became possible to determine the sequence of nucleotides in DNA and thus of the genes that code for the sequence of amino acids in proteins. At the outset DNA sequencing was subject to many errors. For example, missing only one single nucleotide in the many hundreds that make up the gene completely changes the derived amino acid sequence to an erroneous one. However, using our "GCMS" methodology to quickly determine a few short amino acid sequences from the corresponding protein, we could readily detect and correct such errors with relatively little effort. Thus in the late 1970s and early 1980s we determined, in collaboration with a number of DNA sequencing laboratories, the structure of quite a few so-called "aminoacyl-tRNA-synthetases", proteins 500 to almost 1000 amino acids long. They are enzymes which play a very important role in the biosynthesis of other proteins in the cell, from bacteria to human.

Over the past two decades, instrumentation for mass spectrometry has greatly advanced. In combination with the now much more reliable DNA sequencing, which led to the delineation of the entire human genome, and fast computational methods, the identification of proteins in biological systems has become fast and routine. Because the proteins originally produced in the cell along the genes must be "activated" by shortening and/or chemical modification of certain amino acids, their final number is much greater than that of the corresponding genes. These structurally modified proteins are nowadays identified and characterized almost exclusively by mass spectrometric techniques. The studies of the detailed structure of the proteins in a cell and their biological significance now have even their own name: "proteomics".

7.2.9 *Moon and Mars*

Shortly after we had acquired the first high-resolution mass spectrometer, one of my students used it to examine the organic compounds in meteorites. This got me involved with NASA's Apollo Project and we then analyzed the material the astronauts had brought back from the Moon in July of 1969. It was necessary to demonstrate that there were no harmful microorganisms present in the "moon rocks" before they could be distributed to the scientific community for detailed study. We had to install a mass spectrometer behind a "biological barrier" in the "Lunar Receiving Laboratory" at the Johnson Space Flight Center in Houston (Texas) to carry out these experiments. Fortunately there was nothing harmful to be found.

Shortly before that time we had developed the connection of the gas chromatograph with a mass spectrometer (see Section 7.2.7) and had written extensive computer programs to efficiently record and utilize the resulting large volume of data. This was also of interest to the space agency and I was asked to lead a team of scientists to search for organic compounds on the planet Mars using a miniaturized GCMS instrument. Although planets were very much outside of my field of research here on Earth, it was an interesting challenge. I felt that if it was to be done at all, it should be done well and agreed to take on the task of sending such an instrument (Figure 7.7).

This work on the "Viking" project began in 1969 and lasted through the successful mission in the summer of 1976. After ten months of interplanetary flight, two instruments landed on July 20 and September 3, respectively, safely on the surface of the planet and functioned flawlessly for a few months until shut down because of the approaching Martian winter. A thorough mass spectrometric analysis of the soil, conducted by remote control over a distance of more than 325,000,000 km revealed the absence of organic material on the surface of the planet. To others a "negative" result might have been disappointing and discouraging, but I felt that this finding was just as important. It led to the discovery of the

Fig. 7.7 The author with the flight-spare unit of the gas chromatograph – mass spectrometer sent to Mars. Photograph taken at the Kennedy Space Flight Center, Cape Canaveral, before launch of the Viking spacecraft to Mars in the summer of 1975.

oxidizing characteristics of the Martian surface material and helped in the evaluation of the three biological experiments on Viking, some of which had produced ambiguous results. Having successfully accomplished this "extraterrestrial" assignment, I could again fully concentrate on my work on Earth.

7.2.10 *Epilogue*

It may be gathered from the foregoing that I tried to avoid the main stream and often go my own way. This was possible by keeping eyes and ears open and using my own knowledge and expertise, which offered solutions to problems in other fields or *vice versa*. If I had not, almost by accident, listened to W. H. Stahl's talk at the Chicago conference in 1957, I would never have become enthusiastic about mass spectrometry. Had I not been willing to join the Analytical Chemistry division at MIT, I would probably have become a run-of-the-mill Professor of Organic Chemistry, or have ended up in the pharmaceutical industry.

Going it alone never bothered me. I often heard colleagues asking the Department to hire more people in their specialty so that they have "someone to talk to". I always could talk with any of my colleagues, whether they were organic, inorganic, physical or biochemists. To others in mass spectrometry I could talk anytime on the phone, at small meetings or at international conferences.

Most gratifying was the work with my students and postdoctoral associates whom I trained during these exciting times in a new field, seeing them go out and continue with success either in academia or the pharmaceutical/biotechnology industry. They, in turn, appreciated the harmonious environment in my research group and expressed it in 1996 by creating and funding the "Biemann-Medal". It is awarded annually to

Fig. 7.8 The first recipient of the Biemann-Medal, Prof. Scott McLukey (Purdue University), center; Prof. Veronica Bierbaum (University of Colorado), then President of the American Society for Mass Spectrometry, right; and the author, June 1997.

a young scientist for exceptional achievements early in his or her research career. The award is administered by the American Society for Mass Spectrometry (Figure 7.8).

7.3 Appendix A: Muscopyridine

The Swiss researchers had deduced from physical data (ultraviolet spectrum) that the substance contains a pyridine ring (the nitrogen analog of benzene). Because it was produced by the musk deer, they named it "muscopyridine". Elemental analysis by combustion indicated the presence of 16 carbon atoms, the same number as in an accompanying substance, "muscone", of known structure (VIII). Although it is a very flexible ring, it is shown here in a form that illustrates the close relationship to the proposed structure (I): one just has to replace the O-atom and one H-atom by N to close the ring, and "aromatize" the latter by the removal of four additional H-atoms. Nature does this by placing muscone into a properly folded protein (an enzyme) and brings in a nitrogen atom (probably in the form of ammonia, NH_3) to close the ring. Another set of enzymes then removes a molecule of H_2O and the other hydrogens.

These were the "biogenetic" guesses that led Professor Buechi to propose structure (I) for muscopyridine. But carrying out the process in the laboratory is another matter. Because of the flexibility of the 15-membered ring of muscone (VIII), the chance that the –C=O and the right -CH_2-group come close enough to form a bond are extremely low. Thus he decided that we start out with compound (IX), which is commercially available and much cheaper than mucone. This meant to build up the pyridine ring and to add the -CH_3 group, all of which could be done in the laboratory by using 11 discrete chemical reactions.

7.4 Appendix B: Sarpagine

As illustrated in Figure 7.9, the mass spectra of the two compounds V and VI indeed exhibited the exact same fragmentation pattern, with all peaks shifted by the predicted mass difference (net one O atom = 16 mass units). A similar pair of quite different structure, ibogaine (X) and ibogamine (XI), exhibit mass spectra differing by 30 mass units (net CH_2O), showing a fragmentation pattern very different from that of the sarpagine/ajmaline type. This demonstrates the specificity of mass spectrometry with respect to the structure of such alkaloids. The mass of these fragments can be related to the bond cleavage processes that are involved.

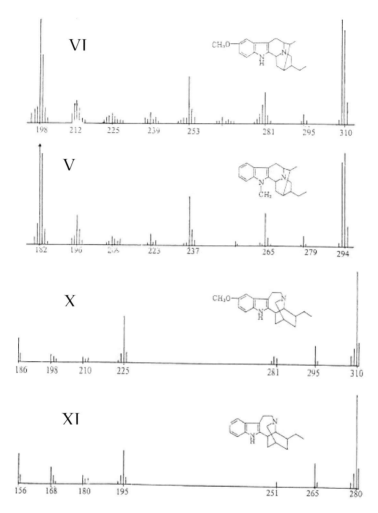

Fig. B.9 The mass spectra (from top) of the conversion product (VI) (molecular weight 310) of sarpagine (IV); the degradation product (V) (mol. wt. 294) of ajmaline; ibogaine (X) (mol. wt. 310) and ibogamine (XI) (mol. wt. 280).

7.4 Appendix C: Peptides and Proteins

The conversion of the involatile peptides (II) to the much more volatile amino alcohols (III) required a little bit of chemistry. To disable the zwitter ion formation, the basic center was eliminated by acetylating the amino group and the acidic one by making the methyl ester of the carboxyl group. I also knew of a chemical reaction, which Paul Karrer (ETH, Zurich) had developed for the conversion of an amido group (-CO-NH-) to an amino group ($-CH_2-NH-$). The former connects amino acids to make peptides, while the latter retains that connection but makes the resulting molecule (III) still more volatile. I also expected that the C-C bond in the newly formed $-NH-CHR-CH_2-NH-$ group would cleave easily in the mass spectrometer, due to the adjacent N-atoms, just as the O-atom in ethanol causes the peak at mass 31 (Figure 7.1) to be so large. Such fragments should provide sequence information.

This was indeed the case, as illustrated in Figure C.10. The large peaks at mass 116, 175 and 260 readily identify the three amino acids and their sequence. Deuterium (D, atomic weight = 2), the heavy isotope of hydrogen (H), was used in the chemical conversion step to eliminate certain mass equivalencies.

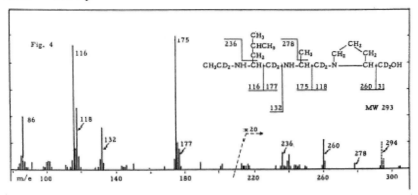

Fig. C.10 The mass spectrum of the triamino alcohol obtained from a tripeptide consisting of the amino acids leucine, alanine and proline, in that order. The peak at mass 294 is due to the attachment of a hydrogen ion to the molecule, a unique property of this compound type. It can be used to recognize the molecular weight of 293.

Chapter 8

James W. Mayer: Pioneer in Semiconductor Device Development

8.1 Introduction by the Editor

8.1.1 *What is a semiconductor device?*

A semiconductor is a solid material that has an electrical conductivity that is between the high values of metals and the low values of insulators. A metal conducts electricity by its electrons. In contrast, a semiconductor has two types of charged species, namely electrons (negatively charged) and holes (positively charged). The presence of electrons and holes allows a semiconductor to exhibit electrical and optical properties that are very different from those of metals. As a result of these special properties, a semiconductor can be used to make electronic devices, such as diodes and transistors. These devices are the heart of electronics (particularly integrated circuits, Fig. 8.1) and are responsible for the electronic revolution that started in the 1950's and culminated in the vibrant electronic industry of today.

The development of a semiconductor device involves the making of high-purity semiconductor materials in single crystal and thin film forms, chemical modification of the semiconductor by doping with foreign atoms, and the fabrication of a device by putting together different types of materials (e.g., semiconductors, metals and insulators) in a microscopic scale. Because of the need for precise control of the material composition and perfection, the fabrication of a semiconductor

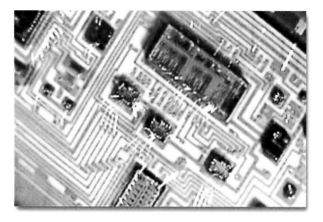

Fig. 8.1 A part of an integrated circuit chip, as viewed under a microscope.

Fig. 8.2 Dr. Mayer at his office.

device involves high vacuum and clean rooms and requires characterization of the materials in a microscopic scale. Device development also involves device design, which requires understanding of how the electrons and holes move in the various materials that make up the device.

8.1.2 *Scientific contribution of Dr. Mayer*

Dr. Mayer (Fig. 8.2) is a pioneer in semiconductor device development. He has developed devices for the measurement of the energy of nuclear particles, ion beam methods for characterizing semiconductor materials, and techniques for fabricating semiconductor devices. In particular, the devices that he developed for the measurement of the energy of nuclear particles have dominated experimental nuclear physics for two decades. Furthermore, Dr. Mayer has contributed much to the application of ion-implantation (a technique of introducing foreign species to a semiconductor) to the semiconductor industry.

Dr. Mayer holds 11 patents. In addition, he is co-author of 8 books and author or co-author of over 700 publications. His books include Backscattering Spectrometry (Academic Press 1978) and Ion Implantation in Semiconductors (Academic Press 1970).

8.1.3 *Educational contribution of Dr. Mayer*

Dr. Mayer was a professor of electrical engineering at the California Institute of Technology (Pasadena, California, USA, 1967-80), an endowed chair professor of materials science and engineering at Cornell University (Ithaca, New York, USA, 1980-92), and an endowed chair professor of science and engineering at Arizona State University (Tempe, Arizona, USA, 1992-present). He has supervised the Ph.D. theses of over 40 graduate students, in addition to teaching an enormous number of undergraduate and graduate students through lecture and laboratory courses. Dr. Mayer loves teaching and is known for how he cares for all his students. He served as the Master of Student Houses at the California Institute of Technology in 1975-80.

8.1.4 *Honors received by Dr. Mayer*

Dr. Mayer is a member of the U.S. National Academy of Engineering (elected in 1984). He received the Von Hippel Award from Materials Research Society (1981), Honorary Degree of Doctor of Science from State University of New York at Albany (1988), Silver Medal from University of Catania, Italy (1986), and Teaching Award from Cornell University (1990). In addition, Dr. Mayer is a Fellow of the American Physical Society and of the Institute of Electrical and Electronic Engineers.

8.1.5 *Career development of Dr. Mayer*

Dr. Mayer received the B.S. degree in Mechanical Engineering from Purdue University (1952) and Ph.D. degree in Physics from the same university in 1960. Prior to becoming a professor, he worked in 1952-54 in the U.S. Army Ordnance Corp. (organized by the U.S. government in 1812 for establishing forces to prepare for the second British War) and in 1959-67 at the Hughes Research Laboratory (abbreviated HRL, created by Howard Hughes[1] around 1950 to address the most challenging technical problems), Malibu, California, USA.

8.2 Dr. Mayer's Description of His Life Experience

My career in research and education had no start at all and was the farthest from my thoughts when I graduated from Purdue University with a B.S. degree in Mechanical Engineering in the Industrial Engineering Option. As an ROTC (abbreviation for Reserve Officer Training Corps, which was established according to the U.S. National Defense Act of 1916 to provide a source of commissioned officers for the U.S. military units) graduate, I was called into active military service and one morning I awoke and declared that I would not spend another day without learning something. So I joined the Army reserve and entered graduate

---

[1] Howard Hughes (1905-1976) of USA was well-known as an aviator, movie producer/director and businessman.

Fig. 8.3 M-48 tank of the U.S. Army (not presently in use).

school with the idea that I would design the production line for the nuclear power packs that would increase the mileage of the M-48 tanks (Fig. 8.3, tank developed by the U.S. Army during the early 1950s for combat in Europe against Soviet tanks, having one of the most advanced fire control systems of that time) I had been working on. Little did I know that such mini-reactors had not been developed and only dreamed of for long-range aircraft. For a summer and two semesters at Purdue University, I tried one department after another. I did take a course in Quantum Mechanics from Professor Hubert M. James (1908-1986) in Physics. I still remember it as the most beautiful course I ever took. It started from the hydrogen atom and developed quantum theory in a clear and logical manner. Unbeknownst to Professor James, he became my inspiration for my teaching (which was still far in the future as I was never a Teaching Assistant at Purdue). I did need to support my wife and children, as the G.I. Bill of Rights (USA law to provide education, training, unemployment assistance and housing to returning veterans) did not go far enough to pay expenses. I replied to an opening to assist the Physics Department Chair in a project at a small industrial company, Duncan Electric. The project was to put a Schottky barrier device on germanium and insert it into a hypodermic needle. At the time I did not know what a Schottky barrier[2] was, much less "germanium". Much later

---

[2] A junction of two materials, commonly a metal and a semiconductor, such that it allows electrical conduction across the junction in only one direction, thereby enabling its use as a rectifying contact.

I learned that the Schottky barrier was a metal film deposited on a semiconductor, germanium in this case. The barrier allowed easy flow of electrical current for one polarity of voltage applied to the semiconductor and blocked flow in the other. The Schottky barrier is one of the simplest rectifying contacts.

I applied for the job and the Physics Department Chair, Professor Karl Lark-Horowitz (1892-1958, Fig. 8.4) accepted me without listening to my claims of ignorance. After all, I did pass the course of Professor James. The project was at a standstill because the Physics postdocs could not get rectifying contacts made at Duncan Electric. I solved the problem by asking the production engineer to go back to the original method of producing contacts. He did and the rectifying contacts worked. I was a hero, much to my amazement, as the solution was straight from my background as an industrial engineer. Lark-Horowitz did encourage me to become a graduate student in Physics and assigned Professor Ben Gossick as my advisor. Gossick suggested that I use the Schottky barrier to measure the energy of alpha particles[3]. I soon learned that alpha particles were energetic (million electron volt energies) helium ions, the helium atom ionized to remove both atomic electrons, emitted from radioactive nuclei such as polonium-210[4]. The electrical signal,

Fig. 8.4 Professor Karl Lark-Horowitz of Purdue University.

---

[3] An alpha particle is a subatomic fragment consisting of 2 protons and 2 neutrons. It is identical to a helium nucleus.
[4] Polonium (atomic number = 84) was the first element discovered by Mme. Curie (native of Poland) in 1898. It is a very rare natural element. Uranium ores contain only about 100 micrograms of the element per ton. Polonium-210 (with atomic mass 210, as the nucleus consists of 126 neutrons and 84 protons) is the most readily available of the 25 isotopes of the polonium. It is an alpha emitter with a half-life of 138.39 days.

produced by the rectifying contact on germanium was small, millivolts and I learned about amplifiers and pulse-height analyzers. I published my first paper in 1955 and received the Miniaturization Award in 1958 for development of a nuclear particle probe mounted in a hypodermic needle for use in brain tumor therapy.

As far as teaching is concerned, I had role models. I was fortunate at the California Institute of Technology because the students were the same age as my own children so that bare-feet, long hair and the look were no surprise. I enjoyed the students and the course material. At the beginning of my teaching career, it often required 8 hours of preparation for each hour of lecture. More recently, I have adopted the case study approach as much as possible so the students learn to present the material. They learn the content if they have to stand in front of the class. After 35 years in the classroom, I still teach 2 to 3 courses a semester. Currently I teach in the Art Department, "The Science of Paintings", in the Physics Dept, "Patterns in Nature" - an internet course with 200 students, and in Materials Engineering, "Electronic Thin Film Science". They are all courses I developed in collaboration with faculty or visiting scientists. I enjoy testing ideas with both colleagues and class members.

My research career is based on the discovery and development of the semiconductor nuclear particle spectrometer (detector) which permits fast accumulation of ion beam analysis data and the interaction with over 100 graduate students and visiting scientists in my 3 University labs:

Fig. 8.5 Werner Van der Weg (Utrecht), James W. Mayer, and Jozsef Guylai (Budapest) at a conference in 1976.

Caltech (short name for California Institute of Technology), Cornell University and Arizona State University. Figure 8.5 gives the flavor of our interactions with Werner Van der Weg (Utrecht, Netherlands), myself and Jozsef Guylai (Budapest, capital of Hungary) at a Conference in 1976.

The detector came out of my Ph.D. thesis at Purdue University where I showed that a simple Schottky barrier on germanium could act as an alpha particle spectrometer. I mounted the detectors in hypodermic needles for work at Massachusetts General Hospital (the third oldest hospital in USA, located in Boston and consistently ranked as one of the country's best hospitals) for their brain tumor (glioblastoma multiform) therapy using neutron irradiation of boron-10 (a stable isotope of boron) containing tissue (the n, alpha reaction). I had turned in my thesis and left for Hughes Aircraft to work on p-n junction silicon detectors when I was asked back to defend my thesis. This came about because Ward Whaling at Caltech had invited me to give an invited talk at the December meeting of the American Physical Society (a learned society that was started in 1899, with over 40,000 members in 2005). The Purdue Physics Department felt I should have some Physics degree as I only had a B.S. in Mechanical Engineering (I had skipped the M.S. in Physics). In one frantic day, I passed my French and German literacy tests, defended my thesis and presented a seminar on my detector work which stretched out for 3 hours.

Back at Hughes I learned to make diffused p-n junction (junction between a p-type semiconductor and an n-type semiconductor) detectors. I made the detectors at the Semiconductor Division of Hughes and then drove back to the Airport site to test their response to alpha particles. My proud moment was when I was told that one of my detectors was used as a laboratory standard at Argonne National Laboratory (the first national laboratory of USA, chartered in 1946).

I was transferred to Hughes Research Laboratory in Malibu and worked on and patented the lithium drift p-i-n (involving a p-type semiconductor, an intrinsic semiconductor and an n-type semiconductor) detector as beta[5] and gamma ray[6] spectrometer. Hughes did not pursue the detector patents and even I saw that there was no interest. Then a marvelous joining of forces occurred. A team was formed of Bob Baron (analytical theory), Howard Dunlap (technician extraordinary), Ogden Marsh (experimental genius) and myself. This group is in Figure 8.6 which shows me at the right with pipe in hand, Ogden Marsh[7] in the background, Bob Baron in the white shirt in front of Ogden with Howard Dunlap sitting next to him in the lab coat.

As a team, the four of us exploited the high resistivity of the silicon used in detectors to study electrical injection of electrons and holes.

Fig. 8.6 Dr. Mayer seated on desk, Bob Baron to the lower left and Howard Dunlap seated to the left of him, Ogden Marsh to the upper left of Dr. Mayer and Gus Mohr to the left of him.

---

[5] A beta particle is a fast moving electron (negatively charged) or positron (positively charged). It can travel a few feet through air and can be stopped with a few sheets of aluminum foil. It is ejected from the nucleus during beta decay, which is a process that unstable atoms can use to become more stable. For example, a neutron converts to a proton by emitting an electron, which is a beta particle. Beta particles are more penetrating than alpha particles, but less so than gamma radiation.
[6] Electromagnetic radiation of very short wavelength in the range from 10^{-6} to 10^{-1} nm.
[7] Ogden Marsh (1929-2003) worked at Hughes Research Labs in Malibu for 25 years as an experimental scientist in applied physics. He was internationally known for his work in crystal growth. He was also a visiting professor at California Institute of Technology for more than 30 years, working with graduate students in physics.

Fig. 8.7 Dr. Albert Rose of RCA.

Single carrier injection was a favorite topic at RCA (Albert Rose[8], Dick Bube[9] and others) and we pioneered double-injection by applying forward bias voltage to the p-i-n detector. We formed solid-state plasma of electrons and holes in the intrinsic region. After proclaiming our discoveries we wrote 3 definitive articles for the Physical Review (a prestigious journal for research papers in physics, with a history dating back to 1893). They sank with barely a trace as we received a total of ten reprint requests. In contrast, a simple paper I wrote on the search for gamma ray detectors made the citation index. Again one of my major topics bit the dust.

At this point in 1965, the Research Lab became interested in ion implantation. They had a Cesium ion engine intended for propulsion in space but just sitting at Malibu. It was huge. Fortunately Gus Mohr, sitting to the right of Ogden Marsh in Fig. 5, built a small ion implantation system that allowed us to implant antimony into silicon. I later learned that Gus had developed this system in the 1940's at Bell Labs (one of the most famous industrial research and development

---

[8] Dr. Albert Rose (1910-1990, Fig. 8.7) worked at RCA for 40 years prior to his retirement. He received the IEEE Edison Medal in 1979 "For basic inventions in television camera tubes and fundamental contributions to the understanding of photoconductivity, insulators, and human and electronic vision."
[9] Dick Bube (Dr. Richard H. Bube) is currently Emeritus Professor of Materials Science and Electrical Engineering at Stanford University.

organizations in the world, with strength in communications technologies, history dating back to 1947 and awards including 6 Nobel Prizes in Physics). They used it to form p-n junctions and then turned to other projects that led to the transistor. The team of Baron, Dunlap, Marsh and myself jumped into ion implantation without characterizing the implantation system which had a vibrating carbon-antimony source. We could not measure the ion current and had no idea what was the implant dose of ions per square cm. All we could tell was that the implanted surface turned "milky" which we assumed was a surface coating of antimony. The color disappeared when the implanted sample was heated to 600°C. (It later turned out that the implanted silicon became amorphous and "milky" in color. The amorphous layer recrystallized at 600°C. This phenomenon turned into the subject of an extensive study called "solid phase epitaxy".)

Two events occurred which made another major change in my research career. Marc Nicolet[10] invited me to join the faculty at Caltech based on my studies of electrical injection in semiconductors. I was so excited that I gave a totally obscure candidate seminar on the solid-state plasma laboratory within the p-i-n structure. I then spent 13 years at Caltech with Nicolet as my mentor and friend. The second event was a call from John Davies at the Chalk River Nuclear Laboratories (a nuclear research and development facility started in 1944 and owned by Atomic Energy of Canada Limited and located on the Ontario side of the Ottawa River, a tributary of which is Chalk River) where I went to visit Dick Fowler on gamma ray detectors. John Davies and I then spent days investigating the deeply penetrating tail in the antimony distribution in our implantation studies at Hughes Reseach Laboratories. John was correct—"no tail"—and then he went on to introduce me to Rutherford backscattering studies at Chalk River. The detector that I had developed was now made commercially by Ortec (a leading supplier of radiation detection and measurement systems) and was used routinely in ion beam studies at Chalk River. Chalk River was also the center of ion implantation studies. My big contribution was to co-author two papers

---

[10] Marc-Aurele Nicolet, currently Emeritus Professor of Electrical Engineering and Applied Physics, California Institute of Technology.

on ion implantation that clarified the mysteries of the Hughes studies: one paper was on the electrical measurements by the Hughes team and the other paper was measurement of lattice location of the implanted dopant atoms on substitutional sites and lattice disorder caused by implantation. It was a grand and exciting time. John Davies, Lennart Erickson (Stockholm, Sweden) and I later wrote a book on our understanding of ion implantation.

I then returned to start work at Caltech and used the venerable ion beam accelerator in the Kellogg Radiation Laboratory of the Physics Department there. Figure 8.8 shows me in the laboratory aligning a silicon sample for channeling[11] studies. This illustrates the beauty of the detector as it allows one person to carry out the experiments. In the time of big, multi-body experiments in nuclear physics, the ion beam experiments were direct with clear results.

There was an international group of us that met at Caltech based on our interests in detectors, ion implantation and ion beam analysis. We did not belong to any one academic discipline and we were free to

Fig. 8.8 Dr. Mayer aligning a silicon specimen for channeling studies.

---

[11] Channeling is a form of Rutherford ion backscattering that involves aligning the ion beam along a crystallographic direction of the material being studied. The technique is mainly for analyzing damage.

organize our own conferences and even founded our own Physical Society, the Bohmische Physical Society (Kaiserlich-Konigliche Bohmische Physikalische Gesellschaft). The conferences still go on every year and the Bohmische Society has over 700 members.

At Caltech, we found that the ion beam analysis tool of Rutherford backscattering (RBS)[12] (Fig. 8.10) with MeV helium ions (alpha particles) was a "depth microscope". We could measure changes in composition with a depth resolution of 20 nm at depths several hundred nanometers below the surface. For example, we could detect a thin silicide layer at the interface between the metal layer and the silicon substrate. At this time there was a demand for an improved metallization technology for integrated circuits. The standard deposited Al film on silicon formed deep Al-Si spikes which could penetrate though the source and drain layers. The controlled interactions of metals such as nickel, palladium, cobalt, chromium and others formed stable silicides without spike formation. At Caltech, we turned to the study of

Fig. 8.9 Ernest Rutherford (1871-1937), inventor of the Rutherford backscattering technique of material analysis.

---

[12] An experimental technique discovered by Ernest Rutherford (Fig. 8.9) (winner of the 1908 Nobel Prize in Chemistry) in 1911. It involves directing helium ions (or alpha particles) at a solid, where the ions are scattered from the atomic nuclei in the solid. The energy spectrum of the backscattered ions is recorded. The spectrum provides information on the elemental composition profile along the depth of the specimen without destructively digging into the specimen.

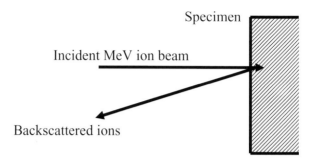

silicides[13]. Cobalt, chromium and others formed stable silicides without spike formation. At Caltech, we turned to the study of silicides[14]. As John Poate[15] of Axcellis, Inc. (a producer of ion implantation equipment) told me recently, "Silicides are with us forever" and there has been a 30-year history of journal publications on the topic.

   More important to my research career, a new breed of scientists began to show up at Caltech. These were the material scientists who would use Rutherford backscattering as a tool for depth microscopy rather than a subject of study in itself. It was an exciting time with new ideas, new people, and new experiments. Figure 8.11 shows Akio Hiraki, one of the new breed, and me having a peaceful smoke at Caltech. We had found that one could achieve the formation of silicon-oxide layers on top of a gold layer on silicon heated to a mere 100°C. Normally oxide formation occurs at 900-1000°C in silicon technology. This astounding result of low temperature oxidation was discovered by accident when we forgot to pump down the sample tube before heat

---

[13] Compounds of silicon, such as nickel silicide, as used in electronic devices to form electrical contacts to silicon.

[14] Compounds of silicon, such as nickel silicide, as used in electronic devices to form electrical contacts to silicon.

[15] Dr. John Poate is the recipient of the 2002 John Bardeen Award from the Minerals, Metals & Materials Society. The award is to recognize an individual who has made outstanding contributions and is a leader in the field of electronic materials. He is the chief technology officer of Axcelis Technologies.

treatment. Akio Hiraki[16] went on to become an Au-Si guru. S.S Lau introduced me to X-ray diffraction and we became close friends.

Fig. 8.11 Akio Hiraki and Dr. Mayer at Caltech.

Fig. 8.12 Dr. Mayer at his desk at Cornell.

---

[16] Dr. Akio Hiraki is currently Professor at Kochi University of Technology, Kochi, Japan.

I had many interactions with the undergraduates at Caltech both as SCUBA instructor for ten years and as Master of Student Houses. I decided that Caltech was destructive of students and was no longer able to recommend Caltech to incoming students. I left Caltech much to the annoyance of my friend Marc Nicolet and set off for Cornell University where I set up another ion beam facility.

I was in the Materials Science and Engineering Department at Cornell and again was surrounded with colleagues, friends, graduate students and visiting scientists. Ed Kramer (now Professor in the Departments of Materials and Chemical Engineering, University of California, Santa Barbara) brought analysis of polymers by forward recoil spectrometry into the arsenal of ion beam techniques. Then, too, I was fortunate enough to be the Ph.D. thesis advisor to four members of the same family: Liang-sun Hung (father; Professor, City University of Hong Kong), Long-ru Zheng (mother; Principal Engineer, City University of Hong Kong), Qi-zhong Hong (son; Texas Instruments Inc., Dallas, TX, USA) and Stella Hong (daughter). They were four of the 27 graduate students who received Ph.D.'s from me at Cornell. Figure 8.12 shows me at my desk at Cornell. There was more paperwork and less hands-on for me. This situation pleased the graduate students.

The cold weather drove me from Ithaca and I relocated to Arizona State University. Again, I set up an ion beam facility and attracted Barry Wilkins from Bellcore to run the facility. Again, there were friends, students and visiting scientists. The thrust of the research remains the same with ion beam analysis serving as a depth microscope. The cycle of collaboration continues: I teach a course with my first graduate student, Tom Picraux[17], at Caltech, and I have a joint research project with two Cornell Ph.D.'s, Mike Nastasi (now at Los Alamos National Laboratory, became Fellow of the Laboratory in 2000) and Terry Alford (now at Department of Chemical, Bio and Materials Engineering, Arizona State University, Tempe, AZ). Of course, I continue collaboration with S.S. Lau (Distinguished Professor, Department of

---

[17] S. Tom Picraux received his Ph.D. degree in Engineering Science from Caltech in 1969. He is now Professor in the Department of Chemical and Materials Engineering, Arizona State University, Tempe, AZ, USA.

Electrical and Computer Engineering, University of California, San Diego) to maintain our 30 year tradition of good times.

My wife, Elizabeth (Betty), has been my support and helpmate throughout my career. It started while I was in the Army Ordnance Corps, 1952 to 1954, stationed in Centerline Michigan at the Ordnance Tank Automotive Command. I worked double shifts in developing and testing tanks. Elizabeth would feed me and other lieutenants who would come to our apartment for poker sessions. She would play with us as she is an avid competitor. In graduate school she provided breakfast on the weekends for the lonely. Then at Caltech, where I was responsible for 1000 students, she prepared meals for 20 students twice a week and fed the SCUBA students after our Sunday dive. Elizabeth also provided rooms for disturbed students when they returned from the hospital. This was an immense help for me in my efforts to combat student depression and suicide attempts (no successful attempts while I was a Master at

Fig. 8.13 Dr. Mayer and his wife, Elizabeth, at Purdue University for the Outstanding Mechanical Engineer Award.

Caltech). At Cornell and Arizona State University, the entertainment program is not as hectic and Elizabeth is always supportive. Nearly all of the postdocs and visiting scientists have spent evenings with us. Elizabeth has made my research a family affair. Fig. 8.13 shows the two of us at Purdue University when I was awarded the Outstanding Mechanical Engineer citation.

Chapter 9

Herbert A. Hauptman: Winner of the 1985 Nobel Prize in Chemistry

9.1 Introduction by the Editor

9.1.1 *Nobel Prize in Chemistry*

Winners of the Nobel Prize in Chemistry over the years include Lord Ernest Rutherford (1908, Great Britain, for investigation of the disintegration of the elements and the chemistry of radioactive substances), Marie Curie (1911, France, for the discovery of the elements radium and polonium), Irving Langmir (1932, U.S.A., for discoveries and investigations in surface chemistry), and Linus Pauling (1954, U.S.A., for research into the nature of the chemical bond and its application to the elucidation of the structure of complex substances). The Prize in 1985 was shared by Herbert Hauptman and Jerome Karle, both of U.S.A., for their work in determining molecular structure, which has been used to develop numerous drugs. In particular, Hauptman and Karle found a method to determine crystal structure, which refers to the periodic arrangement of atoms in a crystalline solid. Furthermore, they worked out equations and procedures for use by scientists to analyze crystal structure through radiation such as x-ray.

9.1.2 *Scientific contributions of Dr. Hauptman*

Dr. Hauptman is a world renowned mathematician who pioneered and developed a mathematical method that has changed the whole field of chemistry and opened a new era in research in the determination of the molecular structures of crystallized materials. Today, Dr. Hauptman's methods, which he has continued to improve and refine, are routinely used to solve complicated structures. The methods have been developed into practical instruments for determining the structure of molecules within both inorganic and organic chemistry – not the least within the chemistry of natural products. In order to understand the nature of chemical bonds, the function of molecules in biological contexts and the mechanism and dynamics of reactions, knowledge of the exact molecular structure is absolutely necessary.

The determination of structure involves generating a three-dimensional picture of the positions of the atoms. The picture maps the electron density within the crystal; the density is greatest at the center of the atoms. It can never be less than zero anywhere, and this is the fact upon which the Hauptman-Karle method is based. Structure determination employs radiation of so short a wavelength that it becomes possible to "see" the atoms. X-rays are normally used for this. This means that the wavelength must be shorter than the distance between the atoms. X-rays striking a crystal are deflected and concentrated in different directions, and the intensities of the deflected rays are measured.

As early as the beginning of the 20th century, chemists possessed a good understanding of the geometrical arrangement of the atoms in carbon compounds of only the smallest structures. But it is only through structure determination using X-ray crystallography that researchers have been able to obtain a detailed picture of the distances between the atoms and of the angles between the various bonds of the larger compounds of primary interest to the life scientists.

Dr. Hauptman has authored over 170 publications, including journal articles, research papers, chapters and books. In 1970, Dr. Hauptman joined the crystallographic group of the Hauptman-Woodward Medical Research Institute (formerly the Medical Foundation of Buffalo) of

Fig. 9.1 Dr. Hauptman with crystal structure models.

which he became Research Director in 1972. He currently serves as President of the Hauptman-Woodward Medical Research Institute as well as Research Professor in the Department of Biophysical Sciences and Adjunct Professor in the Department of Computer Science at the University at Buffalo, State University of New York. Prior to coming to Buffalo, he worked as a mathematician and supervisor in various departments at the Naval Research Laboratory from 1947.

9.1.3 *Honors received by Dr. Hauptman*

In addition to the Nobel Prize, other honors awarded to Dr. Hauptman include election to the U.S. National Academy of Sciences in 1988 and honorary degrees from the University of Maryland in 1985, City University of New York in 1986, University of Parma, Italy, in 1989, D'Youville College, Buffalo in 1989, Bar-Ilan University, Israel, in 1990, Columbia University in 1990, Technical University of Lodz, Poland, in 1992 and Queen's University, Kingston, Canada, in 1993. Additional honors received include Cooke Award, State University of New York, 1987; establishment of the Eccles-Hauptman Student Award,

Fig. 9.2 Dr. Hauptman at the 1985 Nobel Prize in Chemistry award ceremony.

State University of New York, in 1987; Citizen of the Year Award, Buffalo Evening News, 1986; Norton Medal, State University of New York, 1986; Schoellkopf Award, American Chemical Society (Western New York Chapter) 1986; Gold Plate Award, American Academy of Achievement, 1986; the Patterson Award in 1984 given by the American Crystallographic Association; Scientific Research Society of America, Pure Science Award, Naval Research Laboratory, 1959; Belden Prize in Mathematics, City College of New York, 1936; President, Philosophical Society of Washington, 1969-1970; and President of the Association of Independent Research Institutes, 1979-1980.

9.1.4 *Career development of Dr. Hauptman*

Dr. Hauptman received his B.S. degree in Mathematics from City College of New York (New York, New York, U.S.A.) in 1937, his M.A. degree in Mathematics from Columbia University (New York, New York, U.S.A.) in 1939, and his Ph.D. degree in Mathematics from University of Maryland (College Park, Maryland, U.S.A.) in 1955. As a mathematician, Dr. Hauptman is interested in the development of mathematical methods to determine the structure of substances of biological importance.

After more than 20 years with the Naval Research Laboratory in Washingtin, D.C., Dr. Hauptman joined the staff of the Hauptman-

Woodward Medical Research Institute in 1970. He was looking for a fresh venue in which to quietly practice his craft. Then, in 1985, the Royal Swedish Academy of Sciences awarded him the Nobel Prize in Chemistry, changing his life forever. A mathematician by training, Dr. Hauptman would seem to be an unlikely candidate for the Nobel Prize in Chemistry. However, upon further investigation, the reasons for this award become obvious. Although he had taken only one chemistry course in his life, he was able to use classical mathematics to resolve an issue that had stymied chemists for decades.

Around 1950, Dr. Hauptman turned his attention to an interesting puzzle regarding the structure of crystals. Since 1912, chemists had known that a beam of X-rays directed towards a crystal is scattered when it strikes atoms, and the scattered radiation forms a pattern that can be recorded on film. Although the positions of the atoms in the crystal determine the nature of this so-called diffraction pattern, the puzzle for chemists was that they could not readily work backwards from the diffraction data to the atomic arrangement. After perplexing chemists for more than forty years, this problem was finally solved by Dr. Hauptman's mathematical approach. Unfortunately, the procedures, known as "direct methods", that he developed were not immediately understood and appreciated by the chemists who study crystals (crystallographers), and it was many years before he received the recognition he deserved. Today, there are more than 12,000 crystallographers worldwide, and most or all of them use these techniques.

The structures of thousands of molecules have now been solved by crystallographers using Hauptman's direct methods, and many new molecular structures are added to the list each year. As a result of the information obtained in these studies, many new drugs have been designed. Shortly after he received the Nobel Prize, the Buffalo News stated that "Hauptman ... undoubtedly saved more lives ... than anyone else in recent history ... From ... Nobel-winning research in the 1950's have come drugs that combat heart disease and other ailments, and the promise of even more advances in the future."

Dr. Hauptman's current work builds on his earlier Nobel-winning research. He and his colleagues at the Hauptman-Woodward Institute are presently working to extend the methods of structure determination to very large molecules of biological importance, including the proteins that are the targets for drug-design efforts. Indeed, they have achieved new success in recent years by developing a procedure known as "Shake-and-Bake" that has greatly extended the power of direct methods. Currently serving, at the age 88, as president of the research institute that bears his name, Dr. Hauptman does not know the meaning of "slowing down" or, for that matter, "retirement". He continues his work in earnest with the hope that his latest contributions will also have an impact on health care. In a January 1987 article that appeared in *Western New York Magazine* honoring him as "Western New Yorker of the Year", Dr. Hauptman said, "When you look at the great strides that were made against polio and tuberculosis, those breakthroughs could not have been made without research that was done 50 or 100 years earlier … And once in a while, with a little luck, lightning strikes…".

The Hauptman-Woodward Medical Research Institute (HWI) is an independent, non-profit, biomedical research facility located in the heart of downtown Buffalo's medical campus. For almost half a century, HWI scientists have been committed to improving human health through study, at a molecular level, of the causes and potential cures of many diseases. In contrast to clinical research, the focus of Hauptman-Woodward's basic research is to determine the structures of individual substances such as proteins that play a role in the development of specific diseases. This research explores questions like the following: What is the three-dimensional shape of a particular protein molecule? How and with what does this protein interact? What controls these interactions? What structural alterations lead to the development of disease? How can we design drugs having specified therapeutic properties and a minimum of adverse side effects? The answers to these and related questions, often found by Hauftman-Woodward scientists, illustrate the importance of basic scientific research in improving the human condition.

Working under the leadership of Dr. Hauptman, HWI scientists use the techniques of molecular biology, biochemistry, and crystallography to answer these questions. The results of their investigations provide the starting point for better drug design. In addition, other research on-going at HWI seeks to improve the methods of crystallization and data analysis used for molecular structure determination by scientists worldwide.

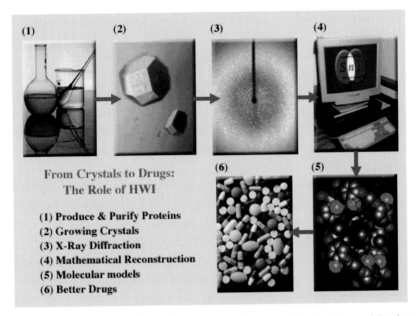

From Crystals to Drugs:
The Role of HWI

(1) Produce & Purify Proteins
(2) Growing Crystals
(3) X-Ray Diffraction
(4) Mathematical Reconstruction
(5) Molecular models
(6) Better Drugs

Fig. 9.3 The research direction of the Hauptman-Woodward Medical Research Institute (HWI)

Fig. 9.4 Dr. Hauptman guiding his staff at HWI.

9.2 Dr. Hauptman's Description of His Life Experience[1]

I was born in New York City on February 14, 1917, the oldest child of Israel Hauptman and Leah Rosenfeld. I have two brothers, Manuel and Robert.

I married Edith Citrynell on November 10, 1940. We have two daughters, Barbara (1947) and Carol (1950).

My interest in most areas of science and mathematics began at an early age, as soon as I had learned to read, and continues to this day. I obtained the B.S. degree in Mathematics from the City College of New York (1937) and the M.A. degree in Mathematics from Columbia University (1939).

After the war I made the decision to obtain an advanced degree and pursue a career in basic scientific research. In furtherance of these goals I commenced a collaboration with Jerome Karle at the Naval Research Laboratory in Washington, D.C. (1947) and at the same time enrolled in the Ph.D. program at the University of Maryland. The collaboration with Dr. Karle proved to be fruitful because his background in physical chemistry and mine in mathematics complemented each other nicely. Not only did this combination enable us to tackle head-on the phase problem of X-ray crystallography, but this work suggested also the topic of my doctoral dissertation, "An N-Dimensional Euclidean Algorithm". By 1954 I had received my Ph.D. degree and Dr. Karle and I had laid the foundations of the direct methods in X-ray crystallography. Our 1953 monograph, "Solution of the Phase Problem I. The Centrosymmetric Crystal", contains the main ideas, the most important of which was the introduction of probabilistic methods, in particular the joint probability distributions of several structure factors, as the essential tool for phase determination. In this monograph we introduced also the concepts of the structure invariants and seminvariants, special linear combinations of the phases, and used them to devise recipes for origin specification in all the centrosymmetric space groups. The extension to the non-centrosymmetric space groups was made some years later. The notion of

---

[1] Partly from the Nobel Prizes 1985, Ed. Wilheim Odelberg, Nobel Foundation, Stockholm, 1986.

the structure invariants and semiinvariants proved to be of particular importance because they also serve to link the observed diffraction intensities with the needed phases of the structure factors.

In 1970 I joined the crystallographic group of the Medical Foundation of Buffalo of which I became Research Director in 1972, replacing Dr. Dorita Norton. My work on the phase problem continues to this day. During the early years of this period I formulated the neighborhood principle and extension concept, the latter independently proposed by Giacovazzo under the term "representation theory". These ideas laid the groundwork for the probabilistic theories of the higher order structure invariants and seminvariants which were further developed during the late seventies by myself and others. During the eighties I initiated work on the problem of combining the traditional techniques of direct methods with isomorphous replacement and anomalous dispersion in the attempt to facilitate the solution of macromolecular crystal structures. This work continues to the present time. More recently I have formulated the phase problem of X-ray crystallography as a minimal principle in the attempt to strengthen the existing direct methods techniques. Together with colleagues Charles Weeks, George DeTitta and others, we have made the initial applications with encouraging results.

I speak for Jerome Karle (the other winner of the 1985 Nobel Prize in Chemistry), as well as myself when I say that our journey to Stockholm to receive the Nobel Prize began some 67 years ago when our parents, with unconscious wisdom gave us a most precious gift, the freedom to grow as we wished, at our own pace, and in the direction of our own choosing. We chose to read a great deal, as soon as we were able, in all areas of science. To their credit our parents permitted, even encouraged, this activity when there may have been moments when they secretly questioned the wisdom of our course. We wish to make grateful acknowledgement of our indebtedness to them for their sacrifices on our behalf.

We are grateful, too, for the opportunity to have attended the City College of New York, at a time when a free education was provided to those who qualified and who would not otherwise have been able to obtain a higher education. Without this splendid gift it is doubtful that we would be here today.

We are also indebted to the Naval Research Laboratory for supporting us in our pursuit of scientific knowledge for its own sake.

We wish finally to thank our wives for their continuing support and encouragement, particularly during the early years when our work was received with some skepticism.

We were fortunate, too, that our particular qualifications, Jerome Karle's in physical chemistry and mine in mathematics, were the exact combination which was needed to enable us to tackle, with some hope of success, the phase problem of X-ray crystallography, the major stumbling-block in the solution of crystal structures by the technique of X-ray diffraction. Our sole motivation was to overcome the challenge which this problem presented, and our satisfactions came from the progress we made. We were fortunate, indeed, that the implications for structural chemistry turned out to be so far reaching; we did not anticipate them.

Chapter 10

Agnes Oberlin: Leader in Analyzing the Structure of Carbon Materials

10.1 Introduction by the Editor

10.1.1 *What are carbon materials?*

Carbon materials (also known as carbons) refer to solid materials that are mainly composed of carbon, which is one of the most important and versatile elements. Carbons include graphite, diamond and fullerenes. Graphite is the most common form of carbon; it is available in natural and synthetic forms.

Synthetic graphite is commonly made from organic materials such as pitch. The process of conversion involves various stages of heating. The process conditions, such as the temperature, the time at each temperature and the gaseous environment, greatly affect the resulting material. Therefore, the selection and control of the process conditions are essential for the making of synthetic graphite.

A synthetic form of graphite that is particularly important is carbon fiber, which is widely used as to reinforce polymers in composite materials for lightweight structures such as aircraft. Carbon fiber composites are attractive for lightweight structures due to their combination of high stiffness, high strength and low density. They are the dominant material for aircraft structures. In contrast, steel is high in density, though it is strong and stiff.

Another synthetic form is graphite film, i.e., graphite in the form of a thin film on a foreign substrate. The film may be of thickness around 1,000 Å, where 1 Å = 10^{-10} m. A graphite film can be used for electrical interconnection, electromagnetic shielding, antistatic surfaces, heat transfer, lubrication, etc.

10.1.2 *What is the structure of carbons?*

The structure of carbons refers to the internal structure in an atomic, molecular or microscopic level. An aspect of the structure is how the carbon atoms in a carbon solid are arranged. In the case of graphite, the carbon atoms are arranged in atomic layers that are stacked up, such that the atoms within each layer are arranged as in a honeycomb (Fig. 10.1). The atomic arrangement in Fig. 10.1 is an example of the orderly and periodic arrangement of atoms in a crystalline material. Within each layer in graphite, the bonding is strong; between the layers the bonding is weak. Therefore, the layers can slide with respect to one another quite easily, thus making graphite a lubricant and a pencil material.

In a carbon fiber, the layers are not flat but are preferentially oriented parallel to the axis of the fiber. This preferential orientation makes the fiber mechanically strong along the fiber axis. The lateral size,

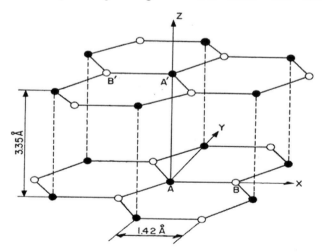

Fig. 10.1 Crystal structure of graphite

thickness, flatness and parallelism of the layers in a stack are all structural attributes that greatly affect the mechanical, electrical and thermal behavior of a carbon fiber. Therefore, knowledge of the structure is essential for understanding the behavior. Understanding the relationship between the structure and the behavior allows appropriate modification of the structure for the purpose of improving certain behavior.

Knowledge on the structure of a material can be gained by experimental methods such as microscopy and diffraction. Diffraction is the main method for determining the periodic pattern of arrangement of atoms in a crystalline material, whereas microscopy is the main method for observing imperfections in the atomic arrangement. The types and amounts of imperfections depend on the process in which the material is made and they greatly affect the properties of the material.

Microscopy conducted using an optical beam is called optical microscopy; that conducted using an electron beam is called electron microscopy. Electron microscopy provides much higher resolution than optical microscopy. Both electron microscopy and electron diffraction can be conducted on the same specimen using a transmission electron microscope (TEM).

10.1.3 *Scientific contributions of Dr. Oberlin*

By using primarily transmission electron microscopy, Dr. Oberlin (Fig. 10.2) has obtained important and detailed knowledge on the structure of carbon fibers. Her extensive research has resulted in a three-dimensional model of the structure of carbon fiber. This model has been used by researchers all over the world for the purpose of understanding the behavior of carbon fibers. In addition, through microscopic observation at various stages of conversion of pitch to graphite, Dr. Oberlin has provided understanding of the process of making graphite from pitch, petroleum products and other organic compounds that contain carbon. The understanding of this process has helped advance industrial processes used for making various graphite products. Furthermore, she has studied the structure of graphite films and porous carbon. The

Fig. 10.2 Dr. Agnes Oberlin

research findings of Dr. Oberlin can be found in her numerous publications, which include 205 papers.

In addition to scientific research, Dr. Oberlin was effective in the direction and supervision of research, as shown by her serving as a director of research in the French National Center for Scientific Research (CNRS) from 1970 until her retirement in 1993. The Center was created in 1939, in the early days of World War II, by decree of President Albert Lebrun. It was the brainchild of a handful of scientists, especially Jean Perrin, who received the Nobel Prize in Physics in 1926. CNRS is the national public organization in France that is devoted to basic research. In addition to directing the research of professional scientists, Dr. Oberlin has directed the doctoral thesis research of 39 people.

10.1.4 *Honors received by Dr. Oberlin*

Dr. Oberlin received the Charles E. Pettinos Award in 1983. This is an international award administered by the American Carbon Society for recognizing outstanding recent research accomplishments in the field of carbon science and technology. She is the first woman in the world to receive this award. Dr. Oberlin has delivered plenary lectures in conferences in USA, Germany, Great Britain, France and Japan. Moreover, she has been honored with professorships in Japan, Switzerland, Mexico and USA.

10.1.5 *Career development of Dr. Oberlin*

Dr. Oberlin was brought up in Paris, France. In spite of limitations in education and research due to the Second World War, she pursued her education in University of Strasbourg located in Clermont-Ferraud, France, where she found interest in crystallography (a topic that was the foundation of her research throughout her career). She completed her degree of Doctor of Physical Sciences in CNRS (Paris) in 1950. Her roughly 50 years of research in CNRS (first in Paris and then in Orleans, France) were dedicated to the analysis of the structure of carbon materials.

10.2 Dr. Oberlin's Description of Her Life Experience

I was brought up as a well educated girl (Fig. 10.3). At the age of 15 in 1940, I learned to resist – resisting the war, resisting an empty future, and resisting to move more and more south in France in order to escape the threats of Second World War.

Due to the War, examination at the end of college (bachelor's degree level) was cancelled in France in 1940. In 1942, even examinations for engineering schools (Grandes Ecoles) were cancelled. The high schools were closed. I entered the University of Strasbourg, which was located in Clermont-Ferrand, a city south of Paris at the center of France (Fig. 10.4). University of Strasbourg had been displaced from Strasbourg (a city east of Paris, at the eastern border of France, near Germany) in 1939 because of the War and the resulting transformation of the area to an army zone, which was later occupied by the Germans. I was ready to work hard in order to pass the four parts of a license examination in two years. The subjects in the examination were mathematics, physics, chemistry and crystallography. Crystallography is the science related to the structure of crystalline materials, which are materials in which the atoms are ordered in a periodic pattern. This pattern is associated with symmetry of various types. Crystallography was my first hobby (ever since 1941). It was a game to play with the 32 types of symmetry, starting only with two of the three symmetry elements, namely plane, axis and center.

Fig. 10.3 Oberlin in Paris, 1933.

Fig. 10.4 Oberlin at the University of Strasbourg in Clermont-Ferrand, France, 1942.

In Clermont-Ferrand, I was at a distance of 76 km from my family in Montlucon (a small town at the center of France). However, it was like being as far as the moon. It took me some luck to have the chance to stand on the step of a car (outside the car, I mean). It was a three-hour journey, due to the resistance fighters who often stopped the train at a tunnel before the gorges of the Sioule River. After arriving safely at my hometown, I often had to stay in the station because of the curfew. This experience makes me realize that the best thing I can advise young people is to resist and follow a single line, which is yours!

In May 1945, I went to Strasbourg (a city east of Paris, at the eastern border of France, near to Germany) with my boss, Professor Hocart, to work toward a diploma in the area of epitaxy (relative orientation between two crystals that are in contact – a concept that is widely employed now in semiconductor engineering). I spent one year in the laboratory to fight with an optical microscope that dated back to the beginning of the 19th century. It was a beautiful design, but...anyway my first publication in the journal Comptes Rendus of the Academy of Sciences was a major event despite the microscope.

A two-year position as a beginning researcher in CNRS (National Center for Scientific Research) was available in Paris at the beginning of 1947 in the group of Marcel Mathieu. I was fortunate to obtain it without knowing him. He was the head of the crystallography group in the Laboratory of Government Chemical Services. This title was chosen for the laboratory during the war to disguise a military organization into a partly industrial one.

M. Mathieu kindly proposed to pick me up upon arrival in Paris, so our first contact was on the platform of the train station. He described himself quite clearly before meeting me; "I am blond haired with long hairs, I am slightly pot bellied and wear systematically a la valliere tie (neck tie with a large bow)..."

He brought me to his laboratory and there I was immediately filled with admiration of the high level scientists present. I was particularly impressed by Rosalind Franklin (1920-1958, Fig. 10.5), who became famous later on due to her pioneering work that led to the discovery of DNA, the genetic material of all living things on Earth. (See the book of Brenda Maddox, "Rosalind Franklin: The Dark Lady of DNA", Harper

Fig. 10.5 Rosalind Franklin (1920-1958).

and Collins Pub., 2003, especially p. 100-103).[1] She was working with J. Mering. She had a strong personality and was a high level scientist and a woman who knew where she wanted to go. Her determination was an example that I tried later on to imitate. She said, "Science and everyday life cannot and should not be separated...by doing our best we shall come nearer to success..."

Coming back to my thesis, the first and almost the only information given on the topic was the statement, "Mademoiselle, you will study the x-ray form factor for microcrystals (i.e., the shape of the reciprocal nodes). For that you have here an x-ray tube with a rotating anode, which will give you diffraction patterns in a short time. You will prepare

---

[1] Rosalind Franklin applied her chemist's expertise to the unwieldy DNA molecule. After complicated analysis, she discovered (and was the first to state) that the sugar-phosphate backbone of DNA lies on the outside of the molecule. She also elucidated the basic helical structure of the molecule. After John Randall (Franklin's supervisor at King's College, England) presented Franklin's data and her unpublished conclusions at a routine seminar, her work was provided – without Randall's knowledge – to her competitors at Cambridge University, Watson and Crick. The scientists used her data and that of other scientists to build their ultimately correct and detailed description of the DNA's structure in 1953. Watson and Crick were awarded the 1962 Nobel Prize in physiology for this discovery. Franklin's career was eventually cut short by illness. It is a tremendous shame that Franklin did not receive due credit for her essential role in this discovery, either during her lifetime or after her untimely death at age 37 due to cancer.

a series of suspensions, using particles of various sizes. Try to start with a particle size of 500 Å and stay in the range from 500 down to 50 Å. Ten samples may be enough." After that, he gave me a book of the size of a dictionary and written in German. (Owing to God, my father understood German and also loved me.) I did not know what was worse between microcrystal preparation using the one hundred recipes in the German book and the maintenance of the x-ray tube of the x-ray diffraction system.

At that time x-ray tubes were some kind of gadget and this one was particularly vicious. The anode was a hollow disc of copper, 50 cm in diameter, refrigerated by water running inside and having the vacuum in the tube controlled from the outside. To allow rotation while maintaining the vacuum in the tube, a conical seal placed in a conical hole was turning around its axis with a lot of lubricating grease.

The scenario was absolutely repeatable: the disc began to turn and the x-ray beam began to heat the copper and the metallic cone, the grease melted and flowed, the seal got stuck, beam heating made a hole in the disc, water went into the vacuum … so Agnes dismantled the tube and cleaned it with cups and cups of benzene, repaired the disc … the disc began to rotate and x-ray beam to heat the copper … so Agnes …

I resisted almost one year to the combination of microcrystals and x-ray tube. Then I decided that I should have been better inspired to be a machinist than a researcher! I was saved by the Marshall plan, which offered to my laboratory the second transmission electron microscope (TEM) available in Europe (it was EMV 2 A RCA 1948 or 1949). I was responsible for this microscope (Fig. 10.6). I was also saved by success in microcrystals preparation. Thus I completed my Ph.D. thesis in 1951 (Fig. 10.7) after many trips between Paris and Marseille (a city at the southern coast of France) to use powerful generators of ultrasonic waves which were only available in the Marine Laboratory of CRSIM (Center for Industrial and Maritime Research). Finally I learned how important it was to know "where you want to go." It was not a brilliant thesis, but I did it alone following my time – a principle which I still obey today; stay on your line and act.

After completing my thesis, I went to J. Wyart Crystallography Laboratory in Sorbonne University (founded in the 13th century, also

Fig. 10.6 Oberlin in Paris, 1948 or 1949.

Fig. 10.7 Oberlin presenting her Ph.D. thesis in Sorbonne University in 1951.

called University of Paris, located in the center of the famous Latin Quarter of Paris at the Left Bank of Seine River) as assistant professor and then associate professor, but staying in CNRS (Paris) as research assistant and then main researcher. For 15 years, I was happy teaching crystallography and being free to choose my research topic. I was still using new TEM's (Models EMU 2D and EMU 2A of RCA Corporation, and one from JEOL, Inc.). I formed a small group of students. Two of them were faithful enough to remain my good friends today.

I married Michel in February 1953 (Fig. 10.8). Both of us were not conformists, so we associated our love of research with the pleasure of adventure. From 1955, our first travel to Sahara (Africa), to 1960, the last one, we visited out-of-the-way places. In 1955, Tassili des Ajjers (Fig. 10.9) was at the far side of yonder (south of Algiers, which is the capital of Algeria) and 700 km on camels was a challenging journey. In 1960 we became fans of kayak, which finally brought us to a new and exciting adventure: exploration of the wild parts of the Herault valley in the south of France. We bought a ruin 30 km north of Montpellier (a city at the south of France, near the Mediterranean Sea). The ruin had no roof, no window, no door and even no floor covering. Up to Michel's death in 1981, we were rebuilding this ruin (known as Mas Andrieu, Fig. 10.10 and 10.11) by our own hands and I got to the end myself with the help of some friends.

In May 1968 a rebellion broke out across France (Fig. 10.12). It began as a series of student strikes at a number of universities and high schools in Paris. The strikes were preceded by confrontations with university administrators and the police. The de Gaulle (President of

Fig. 10.8 Agnes and Michel Oberlin at their wedding in February 1953.

Fig. 10.9 Tassili des Ajjers, Sahara, Algeria (country at the northern shore of Africa, just south of France), Africa, in 1955.

Fig. 10.10 Mas Andrieu in 1995.

Fig. 10.11 View of the country, ruin below the arrow.

France) administration's attempt to suppress the strikes by police action inflamed the situation and led to street battles with the police in the Latin Quarter, followed by a general strike by students and strikes throughout France by ten million workers - roughly two-thirds of the French workforce. The situation was bad in Paris, but worse around the Sorbonne (my laboratory), which was occupied by students. So we

Fig. 10.12 May 1968 rebellion in France.

decided to leave the building promptly. I used this opportunity to move to Orleans. However, it was difficult, because of the dangerous and numerous riots that CRS (Compagnies Republicaines de Securité - riot control force of the police in France) tried to overcome.

With the help of H. Curien and J. Wyart, I obtained along with Michel our independence with the creation of Laboratory ER131 (équipe de recherche). The Laboratory was well equipped and was in the campus of Orleans University, located in Orleans (a city south of Paris and north of Clermont-Ferrand) (Fig. 10.13). We dismantled my three TEM's by ourselves in May 1968. We cried, because of the police tear gas grenades. We then followed Saint Michel Boulevard with the truck dodging in and out among CRS and demonstrators as we moved from Paris to Orleans. After arriving safely at Orleans, we installed the equipment in the empty huge room in the same manner as Mas Andrieu, i.e. with our own hands. We had many apparatus kindly given to us by J. Wyart (my three TEM's) but no money was given. However we were successful to manage a cozy central room around which were distributed offices and laboratories. This meeting place was used by our group for reading, drinking, working, discussion and enjoying in all manners (Fig. 10.14). Between 1968 and 1986, we formed a team of an almost constant number of people (10 to 15, including technicians). The team conducted carbon research and improved the apparatus (EM 300 and then EM 400 Philips electron microscopes). Joy and friendship also

Fig. 10.13 Orleans campus, May 1958.   Fig. 10.14 Orleans campus meeting room.

Fig. 10.15 Christmas, 1977.   Fig. 10.16 Our technician Bergerolte performing "magic".

Fig. 10.17   Around a transmission electron microscope in a laboratory in 1979. From front to back, left to right:
Row 1: D. Joseph, J. Goma, D. Auguié
Row 2: J.N. Rouzaud, M. Villey, B. Deniaux, G. Terriere, M. Oberlin, A. Oberlin, S. de Fonton, M. Lepan, J. Ayache
Row 3: G. Bergerolle, E.M. Zoo Philips "a newcomer"

progressed (Fig. 10.15 and 10.16). In a group consisting of half boys and half girls, goofers were rare (nevertheless there is one in Fig. 10.17).

In remembrance of my thesis supervisor, we named the laboratory Marcel Mathieu Laboratory. He almost never discussed with me about my experiments, but he gave me irreplaceable culture in philosophy, music and literature.

It was too late for Michel to share with me the Pettinos Award (international award for carbon research), which I received in 1983. I mentioned in my award lecture (held in USA as a part of the Biennial Conference on Carbon) that the basic ideas that brought us some success were his. For 28 years, he was my strength. After his death the laboratory endured a two-year crisis prior to the recovery of our consistency. It was particularly difficult for me to gather the students of Michel, as they were accustomed to more freedom. After having formed the habit, they are now still my friends (Fig. 10.17)...

After a short stay in Pau (a city near the southern border of France, near to Spain) as the head of Research Area 124 of CNRS (with support from DRET, a French military organization), I retired in 1993. With the Emeritus position in CNRS, I became scientific advisor in Du Pont de Nemours (a large chemical company) and then in CEA-DAM (a center for industrial research, with emphasis on military applications). I still maintain my line and work with one of my former students, Sylvie Bonnamy, who works in CNRS (Orleans) and whose role I cannot define. Is she my daughter or my successor in research, or both? I think both.

The best advice that I can give to a young scientist is the one I tried to follow: stay on your line, do not disperse among subjects, however attractive they may be. Even now, in the monographs I write with Sylvie, I stay loyal to the carbon field, knowing that there are many problems that remain to be solved.

In conclusion, in 38 years of research and 25 years of the life at Marcel Mathieu Laboratory, 33 scientists and 5 technicians worked with me. A total of 30 students completed their Ph.D. thesis. On the average it took 3 years to complete a thesis. Among them a little more than 10%

were not serious, but among them only one got unmerited success. At the opposite I am proud to have helped with Michel so many valuable people to develop careers. I am also proud that one of them was recently considered for the Nobel Prize.

Twenty-one of my beloved former students remain in the scientific field (either research or industry). Only 10 are still engaged more or less with carbon research.

I am sorry to be probably less bad in writing scientific reports than in writing about my life. However if this article can help someone, it will be enough for me.

Index